CLEARING THE AIR

The MIT Press's publishing mission benefits from the generosity of our donors, including Diana Chapman Walsh.

CLEARING THE AIR

A Hopeful Guide to Solving Climate Change in 50 Questions and Answers

HANNAH RITCHIE

The MIT Press
Cambridge, Massachusetts
London, England

The MIT Press
Massachusetts Institute of Technology
77 Massachusetts Avenue
Cambridge, MA 02139
mitpress.mit.edu

The MIT Press would like to thank the anonymous peer reviewers who provided comments on drafts of this book. The generous work of academic experts is essential for establishing the authority and quality of our publications. We acknowledge with gratitude the contributions of these otherwise uncredited readers.

This book was set in Adobe Garamond Pro by New Best-set Typesetters Ltd. Printed and bound in the United States of America.

Library of Congress Cataloging-in-Publication Data is available.

ISBN: 978-0-262-05274-0

10 9 8 7 6 5 4 3 2 1

EU Authorised Representative: Easy Access System Europe, Mustamäe tee 50, 10621 Tallinn, Estonia | Email: gpsr.requests@easproject.com

For Maeva

Contents

Part III: Renewable Energy

Part IV: Nuclear Power

Part V: Electric Cars

Part VI: Minerals

Part VII: Heating (and Cooling)

Part VIII: Food

Part IX: Cement, Steel, and Other "Hard-to-Abate" Industries

Part X: Carbon Removal and Solar Geoengineering

INTRODUCTION

Try to navigate climate change from the media, and you'll get a daily dose of whiplash. Fossil fuels are going to destroy society as we know it, but the "green" alternatives are just as bad: exploitative, expensive, and a threat to the wildlife that we're trying to protect. You'll find headlines along the lines of "Clean Energy's Dirty Little Secret" in almost every news outlet, from *The Guardian*, *Wired*, and *The Economist* to *The Wall Street Journal* and Fox News.

We're apparently damned if we do switch to clean energy, and certainly damned if we don't. It's no wonder people are left feeling confused and doomed about how to move forward.

Faced with a seemingly unwinnable set of options, most of us seem to react in one of three ways. Some people *do* make the switch toward more climate-friendly alternatives but feel chronically guilty about it. They constantly question their decisions and never feel like they're doing enough. This is no way to live (as I know from experience). Others become paralyzed: they want to make a change but feel lost in the noise of conflicting information. They're overwhelmed, indecisive, and end up sticking with the status quo (which is usually the least climate-friendly option). And finally, there are the backlashers: those who feel conned and lied to about the risks of climate change, or the solutions we'd need to fix it.

I meet and hear from these groups every week: in my inbox, on social media, from journalists, and audiences at talks and book events. They're very engaged, but they are also very uncertain—and, frankly, often deeply defeated—about our chances of getting out of this mess.

What's most heartbreaking is that their vision of the future is so at odds with the opportunity that's in front of us. We *could* have a world run on clean energy, stopping climate change and giving us more breathable cities. We *could* build cities out of green cement and low-carbon steel, have comfortable homes (not too hot or too cold) and no one living in energy poverty. We *could* feed 10 billion people on a fraction of the land we use today, and without killing tens of billions of animals every year. The future is there, waiting for us to take it—or, rather, build it.

Climate change—and the energy, materials, and food systems that drive it—is a massive but solvable problem. We already have many of the solutions we need. Others are still in the pipeline, but thanks to brilliant, tireless innovators, they are getting closer and closer to takeoff. You wouldn't guess it from public discussions, though. Newspapers, documentaries, podcasts, and commentaries are filled with misconceptions and, in some cases, outright myths that plant deep seeds of doubt about our ability to make progress. They claim solar and wind plants aren't that "clean" when you consider all the fossil fuels that are needed to build them. That we won't have enough minerals to build solar panels and batteries. That, rather than making our cities healthier, electric cars increase air pollution. The list goes on and on.

This book is my attempt to investigate and answer the most important and frequent questions raised by these contradictory and confusing claims. It's a guide to help you navigate the minefield of information, and to help you sort fact from fiction when it comes to the solutions we urgently need. For those who are trying their best but feel they might be doing the "wrong" things, or not doing enough, I hope it brings focus to the biggest actions you can take. For the confused and paralyzed, I hope it brings clarity and inspiration about where we go from here. And while I don't expect to reach all the backlashers—those who have already turned cynical—there are always a few who are curious and open enough to listen. Ever the optimist, I'm happy to try.

The structure is very simple: 50 burning questions and 50 straightforward answers. The answers are long enough to give a fuller picture and avoid the pitfalls of a clickbait headline, but only as long as they need to be to get the essential information across (some questions have an extra "Things to bear in mind" section at the end, which is for those who want to dig a bit deeper).

At the heart of every answer is data. That's the only way to understand these solutions: their effectiveness, how big the trade-offs are, and how they compare to the alternatives. Admittedly, as a data scientist who has been studying these trends for years, I might be slightly biased about data's power. But without it, people can hype up solutions that don't work or exaggerate the downsides of those that do. In this book, I'll try to provide clarity with the best data, used in the right context.

There's a lot of ground to cover: everything from public attitudes and polarization to fossil fuels, renewables, nuclear power, transport, mining, food, and controversial topics like carbon removal and solar geoengineering. What emerges from the 50 answers is that there are no deal-breakers. That's why it's a hopeful guide. We are not only capable of solving climate change but also poised to create a better future for ourselves in the process. To do that, we first need to understand that it's possible, then get serious about building the future we want—a future that the media and public discourse would have you believe is impossible.

When it comes to climate change, speed is crucial. If we're to have a chance of meeting our targets, we need to move quickly. Unfortunately, we're getting stuck on questions already answered by scientists, engineers, and economists many times over. We need to stop going around in circles and start building the solutions we need. This book is my attempt to cut through the noise, so we can quit chatting and get to work.

Before we dig into the details, here are three overarching things to keep in mind.

1. PUSHBACK DOESN'T *ONLY* COME FROM FOSSIL FUEL COMPANIES

It's tempting to pin *all* the blame for misinformation on fossil fuel companies and lobbying groups. Now, I don't doubt for a second that fossil fuel companies have downplayed the urgency of tackling climate change and have falsely criticized clean energy to make us skeptical that a move away from coal, oil, and gas is possible. The health and profits of these companies rely on us doubting that we can move to sustainable energy. I believe they have been a huge obstacle to us making this transition to cleaner energy sources. But the reason we're so confused is not *all* on them.

A staggering volume of misconceptions is published in both right- *and* left-wing media, and by commentators on both sides of the aisle. Yes, much more of it comes from the political right, but even some of the most left-leaning outlets and activists have published content that would make anyone skeptical that our solutions will make things better. In 2019, the documentary maker Michael Moore—a strong lefty and Bernie Sanders supporter—published *Planet of the Humans*, an online film designed to expose the dirty secrets of renewable energy and climate activism. Many viewers would come away from it with the conclusion that renewables are just as polluting and disastrous as fossil fuels, and clean energy is just an industry fueled by corrupt billionaires (with young climate activists propping them up). But the film was filled with misconceptions, myths, and outdated facts. It rhymed off countless flawed points against the environmental credentials of electric cars, solar panels, and wind turbines (many of which we'll look at in this book).

Sure, you can argue that this was just one harmless documentary (and everyone's entitled to air their opinion), but it was widely watched, amassing well over 10 million views on YouTube alone. It received four out of five stars in *The Guardian*, a strong left-leaning newspaper that has climate change extremely high on its editorial agenda (granted some other *Guardian* columnists such as George Monbiot did push back).[1] Many viewers would have found the documentary's arguments convincing and assumed that the cleaner alternatives to fossil fuels are, well, not cleaner at all. As the environment policy researcher Leah Stokes headlined in *Vox*: "Michael Moore produced a film about climate change that's a gift to Big Oil."[2] You can see why people would feel information whiplash: In the morning they read about imminent ice-sheet collapse, and in the afternoon, they're being told that there's no benefits to installing a solar panel on their roof, or moving from a gasoline car to an electric one. In other words: The situation we're in is hopeless.

Many climate and clean energy experts did publish fact-checks and corrections. But as "Brandolini's Law" states: "The amount of energy needed to refute bullshit is an order of magnitude larger than to produce it." More people probably watched the documentary than took the time to read the detailed rebuttals and fact-checks, and the damage was already done—an imbalance that is often true of misinformation. That's the issue we're facing: The cost of publishing poor information is incredibly cheap, which means it's almost impossible to keep up with the sheer volume that's published every day. Experts—who usually have full-time jobs so can only try to clean up the mess in their spare time—don't stand a chance.

I don't mean to pick on Michael Moore or *The Guardian*. There are countless other examples that I could have chosen. It's precisely because many of these arguments are coming from those that *do* take climate change seriously that this example makes my point clearly: Politicians, investors, and the public are being fed conflicting information from all angles. It's everywhere.

You can even find examples of pushback from groups who have dedicated their lives to environmental protection. Nuclear power has faced constant opposition, which in some countries has led to the continued burning of fossil fuels (with terrible impacts for climate change and deaths from air pollution). Even some solar and wind power projects have faced protests from environmental NGOs and green political parties.[3]

This comes from a growing tension between traditional environmentalism and what needs to be done to solve climate change. Until recently, environmentalism has been about "stopping stuff": new coal plants, oil pipelines, gas drilling, highways, mines, and infrastructure that encroach on nature. But to tackle climate change, we need to do the opposite: we need to *build* stuff—a lot of it, and we need to do it quickly. This is the dilemma that environmentalists (myself included) are wrestling with: Conservationism has been about blocking and delaying; climate action is about building and accelerating. That's why we sometimes see headlines about green groups opposing clean energy projects. They want to protect some of the landscapes that might be impacted by power pylons and solar farms, but this can come at the cost of more climate change. As will be seen later in the book, I think we can responsibly build clean energy infrastructure, using less land than most people expect.

To be clear, I'm not saying that this is the reason we're in this mess. Environmental activists have played a crucial role in bringing many of the positive changes and progress that we've seen over the last few decades. We've got a lot to thank them for. But when the public sees the strongest and most vocal environmental groups blocking renewable energy projects, it's not surprising that people start having doubts: "If the greens are opposing it, then maybe solar farms *do* need too much land."

So many people still get the basics wrong: More than half of Americans think electric cars are worse for the climate than gasoline, most Brits don't know that nuclear energy is low carbon, and almost everywhere, people underestimate how many others in their country support more climate action (especially among conservatives).[4] Faced with so much conflicting information, it's easy to see why.

Another reason why I think many people struggle to get clear answers is because the answers *aren't* always clear. At least, they're not intuitive. Confusion about climate solutions is not a case of people being dumb or gullible. It's not obvious that heat pumps running on gas need less gas than a boiler, or a world of vegans would need less cropland, not more. It's not even obvious that a low-carbon world might need *less* mining. Some of the questions in this book I didn't know the answer to until I crunched the numbers. Many of the doubts that people have stem from some nugget of truth. They're half-truths. That's why they're so convincing. It does take more energy to manufacture an electric car than a gas one. Wind turbines do kill birds. And poorer countries will need to use some more fossil fuels over the next few decades. What's missing is the context we need to understand whether these are problems we can overcome or deal-breakers that mean our path forward is doomed.

2. DON'T LET PERFECT BE THE ENEMY OF GOOD

Another problem is that we seem to be stuck in something I call "perfect solutionism." People seem to expect solutions to climate change that have no downsides, or they've been falsely promised them and then feel short-changed when that false promise is exposed. Unfortunately, perfect climate solutions don't exist. We have good—even great—ones, but many have some environmental or social cost that we need to wrestle with. When I say "clean energy" throughout this book, I really mean "cleaner energy" because no energy source is completely pollution- or impact-free. We need to be honest about these effects and trade-offs. We also need to recognize where a search for perfection will leave us: in a much hotter world, still hooked on fossil fuels, with millions still dying from air pollution. If we want to move forward, we need to let go of this search for silver bullets.

This perspective matters for how we assess the merits of different solutions. Comparing the waste, mining or biodiversity impacts of renewable energy to a world *without* energy is pointless. Yet that's the comparison that many commentators and critics seem to make. What's more useful is to see how solar or wind power compares to the electricity system we have today: one that mostly runs on coal and gas. No, the impacts of solar and wind are not zero. But they are much, *much* better than fossil fuels.

The notion that one day all our problems will be solved is attractive, but it's an illusion. This is not just true of climate but of social, economic, and environmental problems more broadly. Each generation solves problems and creates new ones for those who come after them. Antibiotics have saved hundreds of millions of lives (at least), but there are now concerns about antibiotic resistance. Would I want to go back in time and undo the miraculous discovery of penicillin and other antibiotics? Of course not. Yes, we have a new problem to contend with, but it's much smaller than the previous one. This is how progress happens—incrementally, from one generation to the next. But it's a journey with no finish line; humanity will never be problem free. As the physicist David Deutsch explains in his Three Laws of the Human Condition:

1. Problems are inevitable.
2. Problems are solvable.
3. Solutions create new problems, which must be solved in their turn.

So, yes, in solving climate change, we will hand down some new problems that our children or grandchildren will have to deal with. They might need to figure out a new form of mineral recycling. Or they'll have to manage small amounts of nuclear waste that we've buried. Do you think they'd rather we handed them the challenge of figuring out how to recycle copper cables, or unchecked climate change leading to a 4°C warmer world with struggling crops and unbearable heat waves? The point is not to deny Deutsch's point 3—that climate solutions will inevitably create new problems—but to see them for what they are: smaller, solvable problems rather than massive barriers that block us from moving forward. Instead of throwing our hands up in the air in despair, we need to roll our sleeves up and figure out how to tackle the problems.

There *are* things we can do to reduce the impacts of renewable energy, electric cars, meat substitutes, and the other things needed to tackle climate change. But we'll have to learn to walk and chew gum at the same time. We need to deploy these technologies, invest money, and form policies while we work on making them better. We simply don't have time to do one after the other.

3. WHAT YOU DO MATTERS: THE FALSE DICHOTOMY OF SYSTEMIC AND INDIVIDUAL CHANGE

We will never achieve perfection, but we can always strive to make things better. That's why I've included a section in every question called "What we need to do." This provides a list of the actions we can take to overcome barriers that are currently getting in the way, *and* things we can do to make this transition as socially and environmentally responsible as we can. These lists include a mix of actions that individuals can take and those that need to be taken by governments, companies, media, or broader society.

When I speak to people who are deeply concerned about climate change and want to make a difference, the feeling that comes through most is helplessness. There's a sense that what they do is pointless. But to say that you or I can do nothing to change things around us is a lie. No, neither of us is going to transform the world or fix climate change on our own—let's not be delusional—but how we think and talk about these problems matters.

Solving climate change is often framed as *either* a systemic problem that governments, big business, and the financial system need to solve, or one that will be solved through individual behavior change (if everyone "just did their bit" we could fix this). But both matter, because they feed into one another. You are part of the system, not separate from it.

Let's take a simple example of shifting the world from gas-powered cars to electric ones or public transport. You could argue that this is a systemic problem: It's on governments to make sure that there are enough public charging points, bus routes, and train lines; it's on manufacturers to build great electric vehicles; and some combination of the two to make both options cheaper than driving a gas car. Those points are all true. But governments and companies are never going to step up unless they know that the public is also on board. There's no point in setting up an electric charging network if individuals refuse to give up their gas cars. No point investing in public transport if no one wants to take the bus. In 2024, the Scottish government ran a trial where it cut the prices of rail journeys during peak hours to incentivize more commuters to take the train. The problem? Not enough people made the switch, and it wasn't financially viable. Positive systemic change can't happen in a vacuum; it needs a positive response from individuals too.

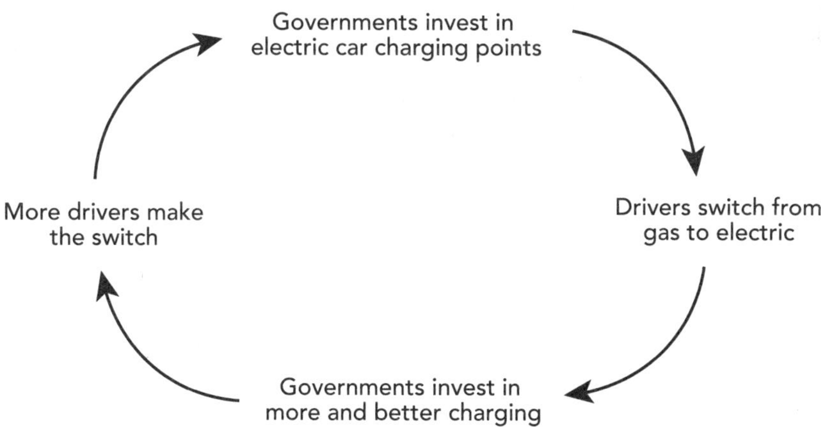

Figure 0.1
Systemic change is reinforced by individual changes: It's not one or the other.

This means that what you do on a personal level—whether it's eating less meat, taking public transport, getting rid of your gas car, or ditching the gas boiler—matters. But so does your support for the large system-level changes that we'll need. Systemic change is driven by culture and public sentiment, and how we all think and talk about climate solutions shapes that culture. So, no, you and I won't be able to fix human rights abuses in mineral supply chains or ensure that nuclear waste is stored and buried safely, but knowing what *can* and needs to be done at a societal level is crucially important for us as individuals. So, yes, in the "What we need to do" sections of each question, focus on the changes that you can make in your own life, but don't dismiss the systemic changes we need as things that "someone else needs to think about." You and I need to understand them so we can push for the right changes too.

I

THE BIG QUESTIONS

We're going to start by tackling a bunch of questions that suggest even *trying* to fix climate change is a total nonstarter. If we can't overcome them, then discussions about specific climate solutions—whether we have enough minerals, or whether wind turbines are bad for birds—are irrelevant.

These doubts are framed around the idea that it's too late to do anything, that change is impossible, or that the efforts of individual countries are pointless. They're often successful in making us feel like the future is bleak and we're powerless to do anything. I've felt that way before. But thankfully these doubts can be overcome.

QUESTION 1

Isn't it too late? Aren't we heading for a 5 or 6°C warmer world?

Answer: Every tenth of a degree matters. There's no point at which it's too late to limit warming and reduce damage from climate change.

I read Mark Lynas's book *Six Degrees: Our Future on a Hotter Planet* when I was 14 years old, and it scared the life out of me. Lynas takes the reader on a journey of what to expect from a world that's 1 degree warmer, 2 degrees, 3 degrees, all the way up to 6 degrees. By the middle of the book, your blood pressure is high; by the end you're on the floor.

It is a well-researched book that offers us a window into many possible futures. Fortunately, the scientific consensus has moved away from the most extreme scenarios since its publication. Unfortunately, a lot of the public messaging has not. Many people believe a pathway to 5°C or 6°C is already locked in, and the only thing we can do now is prepare for the worst.

Let's look at what the latest science says about where we might end up by 2100.[1]

- **Current policies.** This is where we would end up based on the climate and energy policies that governments *have already put in place*. If no countries stepped up their efforts and we continued with current enacted policies, we'd end up at 2.5°C to 3°C higher than preindustrial temperatures by the end of the century.
- **2030 commitments.** If countries met the targets they set for 2030 but enacted no policies afterward, we'd end up at 2.4°C.

- **Net-zero pledges.** Many countries have set ambitious targets to reach "net-zero" emissions—most by the middle of this century. If they achieved this, we'd be in a 1.8°C warmer world.

This is both good news and bad news.

The good news is that we're no longer heading for the worst-case scenarios that scared me as a teenager. The plunging costs of solar, wind, batteries, and electric vehicles, a step up in national policies, and a better understanding of what our energy future might look like have taken us off that terrifying path. And, importantly, countries have put commitments on the table that would keep us "well below 2°C." Now, we'd be naive to assume that they'll all deliver. But it does give us concrete pledges that we can hold governments to account on.

The bad news is that one of our global targets of keeping temperatures below 1.5°C of warming is now out of reach. And a 2.5°C warmer world—which we're on course for—is still a scary and unacceptable one. It could spell the end of many coral reefs. It could cause significant damage to food production, especially in some of the poorest countries. Large parts of the world will experience grueling heatwaves. Arctic sea ice will be gone in the summer. Ice sheets are at a much higher risk of becoming unstable. We really want to avoid ending up there. And we can: 2.5°C or 3°C is not "locked in." There is still time to put ourselves on a better trajectory.

WHAT WE NEED TO DO

First, we can talk and think about our climate trajectory and targets in a more helpful way.

- **Be honest about where we're heading.** The 1.5°C target is dead. If our carbon emissions dropped to zero tomorrow, we *could* achieve it. But the reality is that our emissions aren't going to fall quickly enough (I'm optimistic but I'm not delusional). We need to be honest about this for a couple of reasons. One, countries need to adapt to the post-1.5°C world we'll soon be living in; pretending this is not going to happen robs them of the time they need to prepare. Two, the public—who are repeatedly told that 1.5°C is still within reach—will start to lose trust when we pass that target.

- **Don't throw in the towel.** That's the most important message here: There is no point of no return that makes it pointless to act. Our 1.5°C and 2°C targets are not cliffs or thresholds. Every tenth of a degree is worth fighting for as it reduces the impacts of climate change and limits the damage that's to come. 1.7°C is better than 1.9°C, which is better than 2.1°C. Stop obsessing over arbitrary targets and focus on how you can help reduce our carbon emissions as quickly as possible (see below).
- **Watch out for headlines based on worst-case scenarios.** We're not on the same trajectory to 4 or 5°C that we thought we were a decade ago. Unfortunately, a lot of reporting and studies are still based on these worst-case scenarios. It's sometimes hard for nonexperts to know what scenario is being assumed without reading jargon-filled academic papers. My one quick piece of advice is to look out for any mention of "RCP8.5": this is the acronym of the worst-case (but now implausible) scenario that has often been used in climate modeling.[2] Of course, knowing the impacts of these extreme cases is useful for scientists, but not for policymakers or the public, who assume that this is the most likely outcome. It does make for a great apocalyptic headline, though.

Second, here are the most effective things you can do to reduce your *personal* carbon footprint (many of these—and the questions you might have about them—are covered later in the book).

- If you drive, then cycle, walk, or take public transport more. If you *need* a car, then an electric one is much better than gasoline or diesel.
- If you fly, this will be a big chunk of your footprint. I won't tell anyone to stop flying completely (because for most people, it's not going to happen) but reducing the amount you fly would make a massive difference.
- At home, heating and air conditioning will be your biggest energy-guzzler. Getting your home insulated, and switching a gas boiler for an electric heat pump will slash your home's footprint (and your bills).

- Installing solar panels at home also will reduce your carbon footprint while cutting your energy bills. Some can't afford the upfront costs—or rent a flat where they don't have the option of putting up solar panels—but it's a worthwhile investment for those who can.
- Switch to a renewable energy provider. This also sends a signal that more and more people care about climate change and want low-carbon energy.
- Eat less meat and dairy and move to a more plant-based diet. This doesn't mean you have to go fully vegan; for many, the all-or-nothing approach is daunting. But you can still have an impact by cutting back, especially on beef and lamb.
- Stress less about the small stuff—recycling, plastic bags, and food wrappers, food miles, turning the lights off, leaving devices on standby—especially if it comes at the expense of the big things that are listed above. This is a concept called "moral licensing" where people feel like they've contributed with the small stuff, and therefore ignore their carbon-intensive behaviors. People will often feel proud about bringing their plastic bag to a supermarket (which has a tiny carbon footprint) and then fill it with meat and dairy (which has a much bigger impact).

These individual actions on their own are not going to get us there, of course. At a societal level, we need to go bigger and faster.

- Deploy low-carbon electricity sources like solar, wind, nuclear, and geothermal as quickly as possible. To do that, we'll need massive reforms around infrastructure projects so they can be completed quicker.
- Accelerate advancements in batteries: These technologies hold the key to the energy transition.
- Electrify as many sectors as we can: road transport, heating, steel manufacturing, short-haul aviation. This is the most efficient way to decarbonize.
- Reduce global meat and dairy consumption; innovate on high-quality protein alternatives.
- Invest in forest and ecosystem restoration to suck up lots of carbon.

- Continue innovating in sectors that are not yet ready for large-scale deployment: cement and steel manufacturing, long-haul aviation, and ways to remove carbon dioxide from the atmosphere.

THINGS TO BEAR IN MIND

One of the most common questions I get asked is: Won't we reach a point where it's game over and our planet's systems collapse, a point where we trigger runaway warming?

Tipping points—a threshold where a system moves into an irreversible state—matter, but they don't change what we need to do now to reduce emissions. There are a few key misconceptions about tipping points that are worth looking at here.[3]

First, people often think that the planet has one single tipping point. Or they assume that the 1.5°C or 2°C targets themselves are a global tipping point: that once we pass them, we're thrown into oblivion. That's not true. There's nothing special about 1.5°C. Things are not fine at 1.49°C but disastrous at 1.51°C.

Rather than there being a single *global* tipping point, there are a range of local or regional systems with different ones. Tropical coral reefs are one. The Amazon rainforest is another. The Greenland ice sheet. The Antarctic ice sheet. They won't all "tip" irreversibly at once. While scientists don't know exactly what temperature would trigger these individual points, there is a real risk of doing so, especially as warming gets toward 2°C. We shouldn't hide from the devastating impacts this would have on regional ecosystems. But it's *not* the case that they will set off runaway global warming, taking us to 5°C. Some tipping points will increase global temperatures a bit, but not by whole degrees. For example, if we were to have sea-ice-free summers in the Arctic (which seems likely)[4] this would increase global temperatures by around 0.15°C.[5] A tipping point in the Amazon might have a similar effect. Hitting several of them could increase temperatures by 0.3°C or 0.4°C. That's a lot. But it's not the same as an abrupt change to a "Hothouse Earth."

Another misconception is that these tipping points happen quickly: that if the Greenland ice sheet collapsed, sea levels would rise by 10 meters *within years*. Most of these large tipping points—like ice sheets—play out over centuries or even millennia. It might be 2500 or later before the ice sheet is mostly gone. Now, that would still be terrible—we don't want to hand that problem to future generations. But it's a very different problem from our coastlines shrinking within a decade, which is what people assume when they think about ice sheets "collapsing."

We can still stop this by reducing our emissions and keeping temperatures as low as we can. And almost all scientists agree that it's never "too late" to prevent more damage.

QUESTION 2

Is there enough public support to tackle climate change?

Answer: Not *everyone* cares about climate change, but the majority in every country does.

More people care about climate change than you think. A survey of 59,000 people across 63 countries found that 86% thought that humans were causing climate change and that it was a serious threat to humanity.[1] Nearly three-quarters supported climate policies including carbon taxes on fossil fuels, expanding public transport, more renewable energy, more electric car chargers, taxes on airlines, and protecting forests.

In the chart, you can see the results of another survey across 130,000 people in 125 countries. Eighty-nine percent wanted to see more political action and 86% thought people in their country "should try to fight global warming."

This is even true in "climate-skeptical" America. In that same survey, 74% of Americans thought that the US should be involved in international action to tackle climate change. To make sure that this wasn't just a single fluke study, I looked at more surveys; the results were similar. Sixty-seven percent of Americans thought that developing alternative energy sources was more important than expanding fossil fuels.[2] Seventy-one percent said that climate change is already causing harm to people in the US today.

Support is higher in Europe. A survey in the European Union found that 93% of people believe that climate change is a serious problem.[3] More than two-thirds think that their government is not doing enough.

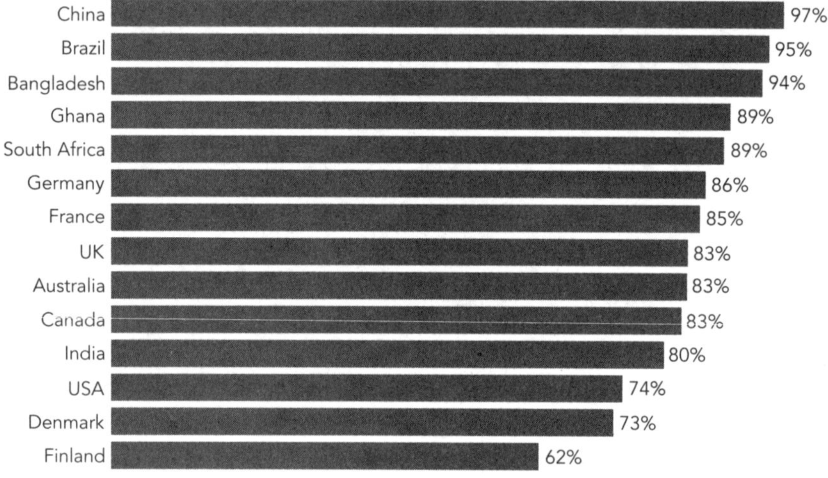

Figure 2.1
The percentage of people who say their government should do more to tackle climate change. *Data source*: Peter Andre et al., 2024.

Studies also find that people nearly always *underestimate* how many others care. In the US, people guessed that just over one-third of their fellow Americans are supportive of climate action.[4] In reality, that figure was twice as high. People thought that those in favor of climate action were in the minority, but they're really in the majority.

WHAT WE NEED TO DO

This perception gap matters because the public appetite for climate policies shapes government action. "There just isn't enough public support on climate change" is the excuse leaders in politics and business use to justify inaction. As individuals, it also makes us more self-conscious to talk about our concerns about climate change, or share what we're doing personally to change things.

What can we do to bridge this gap?

- **Read less mainstream and social media.** Both run on controversy and division. They give us the illusion that for every person who "believes" in climate change, there is another who thinks it's nonsense. Contrarian takes get a platform. If you want to keep up with the news, I'd recommend lower-frequency digests—such as weekly roundups—from outlets you trust. This often filters out a lot of the noise and focuses our attention on the biggest stories going on in the world. I like *The Economist* and a less well-known outlet called Semafor for this.

- **Speak to real people.** This lets others know we care, and it makes us more comfortable talking about climate change too.
- **Focus on solutions, not doomsday headlines.** I speak to a lot of journalists, so this recommendation is for them. People already know that climate change is a serious problem; it's the solutions they're skeptical about. Only writing about the existence or *impacts* of climate change—the wildfires, floods, and hurricanes—is not going to make people less skeptical about the effectiveness or price of electric cars or wind turbines.
- **Lead by example by taking climate actions yourself.** When your whole neighborhood is installing heat pumps, getting rid of gas cars, and installing solar panels, it's hard to argue that "nobody cares." Local behaviors are infectious.

THINGS TO BEAR IN MIND

While most people think climate change is important and are worried about it, it never ranks as the *top* issue they care about.

In polling in the US, around two-thirds of respondents say that climate change is a very or moderately big problem facing America.[5] But it never comes close to topping the list of pressing problems. What always ranks above it? The economy. The cost of living. Health care. Immigration.

So, yes, the public cares about the environment, but most care about the economy and finances more. Human psychology favors short-term problems over long-term ones. This matters for how we communicate about climate action. If climate action is framed as a "sacrifice," then support will fall away. Whether we like it or not, most people are not willing to sacrifice the economy to reduce emissions. And the good news is that they don't have to: Later in this book, we'll see that these two issues are no longer incompatible.

QUESTION 3

Isn't climate action too polarized and politically divisive to fix?

Answer: We're much less polarized than we think, and support for clean energy is strong on both sides of the political aisle.

Many people think that only those on the political left care about climate change. That doesn't make sense when we look at figure 3.1. The US is one of the most politically polarized countries in the world. Climate is a divisive issue. Yet the five American states with the highest share of wind power are all Republican. Red states generate 70% of the country's wind power.[1] The state with the most solar power? Not Democratic California, but Republican Texas.

People overestimate how polarized support for climate action is. Even Republicans themselves underestimate how many other Republicans care about climate change.[2] They also assume that people would *only* support clean energy for climate reasons.

Support for renewables is strong across the political spectrum but Republicans and Democrats support these technologies for different reasons.[3] Democrats support renewables to tackle climate change and air pollution. Republicans back them to reduce energy costs and increase energy independence. Many red states are situated in the so-called wind belt, which is ideal for producing lots of power. The economics work out well for landowners; farmers can make extra cash while continuing to grow crops and raise livestock between the turbines. We might shake our heads and argue that climate change should be higher on their priority list. But if clean energy gets built and we move toward a low-carbon economy, then who cares? The climate certainly doesn't.

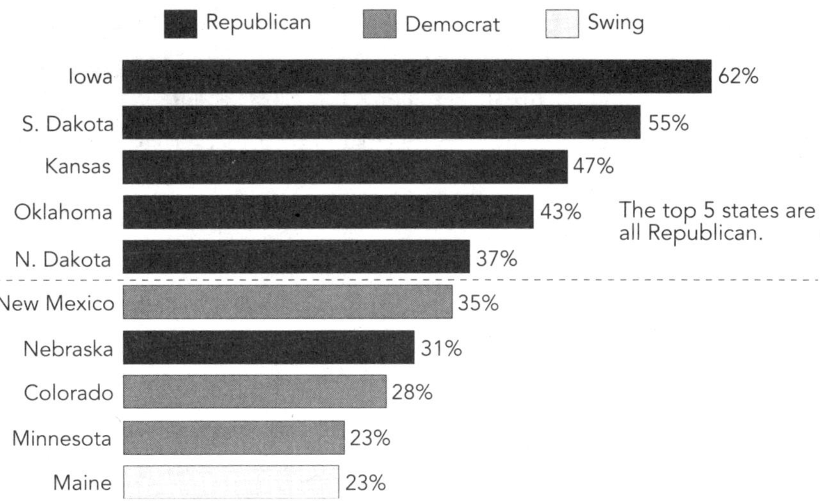

Figure 3.1

The share of electricity that comes from wind in US states.

Data source: US Energy Information Administration. Data is for 2022.

The US has the largest partisan gap—four times bigger than most other countries—so if it can bridge the divide, the rest of the world certainly can. A study from the nonprofit Potential Energy Coalition surveyed nearly 60,000 people in 23 countries to understand public perceptions of climate change, and what messaging is most effective to increase support.[4] Its headline message? "Across most countries, political polarization is not the defining dynamic."

Those who identified as right-leaning in the 23 countries were, on average, 13 percentage points less likely to support climate action. But this was by no means universal: In some countries, like India, conservative voters were *more likely* to be supportive. The partisan gap in most countries was small. And just six of the 83 major political parties didn't have a majority backing for climate solutions. The majority—on both the left and the right—want to see more being done.

To be clear, I'm not saying that partisan gaps don't exist, or don't matter. That would be dumb. Who's in office *does* matter. Going to the polling station and voting for a climate-conscious candidate is one of the most impactful things we can do. But we need to make climate solutions as non- or bipartisan as we can. Over the next 30 years, almost 100 democracies are going to swing back and forth between the political left and right (mostly for reasons unrelated to climate change). Make climate change a "left-only" issue, and we'll be making progress just half of the time, if that. Sorry, that's not going to cut it.

Of course, it doesn't have to be that way: Plenty of countries have support for climate action on the right and left. Making climate change more nonpartisan would help keep our progress steady and consistent, rather than taking two steps forward and one or two steps back every political cycle. Instead of trying to stoke political tensions (which I've seen people on both sides do), we should try to reach across the aisle.

WHAT WE NEED TO DO

How do we make climate action even less polarized?

- **Avoid "crisis framings."** Talking about climate change is good, but messaging around future climate "breakdown" or doomsday scenarios can cause a backlash.[5] Telling conservatives that the end of the world is coming is not going to work. Instead, we should focus on the impacts on local communities now.
- **Avoid judgmental messengers.** No one wants to be guilt-tripped, but this is especially true for those who are already skeptical. The most counterproductive messengers are those who have a political bent to sell. Communicators who listen to people's concerns and are willing to hear arguments from the other side are much more effective. The best are people from *within* a given community: Young conservatives who are excited about the prospects of clean energy are probably some of the most influential individuals out there.
- **Don't frame solutions as limiting people's options.** People hate being told what they can't do, and messages built around limitations always fail. It's much more productive to focus on what we want more of, not what we don't. "Clean energy," not "ban fossil fuels." "Pollution-free cities," not "end gas and diesel." "Sustainable and nutritious proteins," not "ban meat."

QUESTION 4

My country only emits 1% of the world's emissions; surely it's too small to make a difference?

Answer: The world's "small emitters" make up more than one-third of the world's emissions, enough to significantly turn the dial.

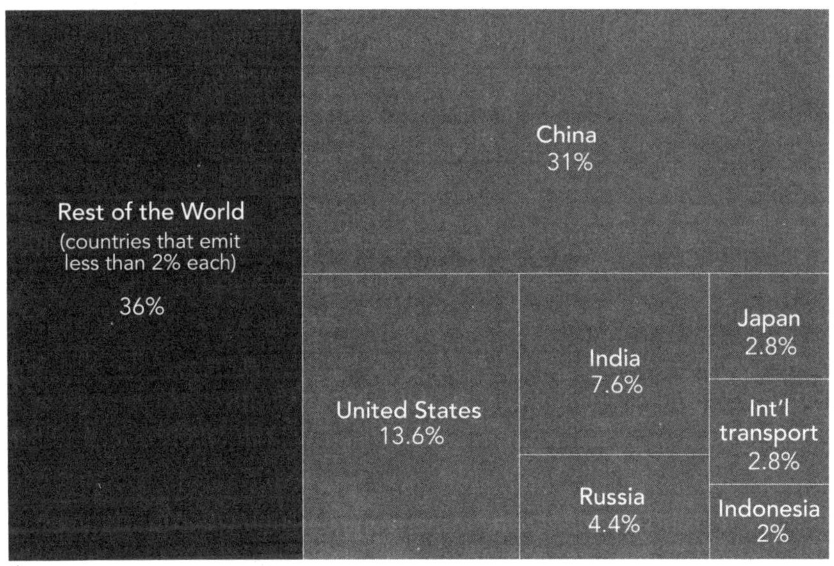

Figure 4.1

Each country's share of global CO_2 emissions.

Data source: Global Carbon Project. Data is for 2023.

If the UK's carbon emissions fell to zero overnight, the climate would barely notice. My country emits less than 1% of the world's carbon dioxide (CO_2) emissions, which is just a drop in the ocean compared to the world's biggest emitters.[1] This gives my government and neighbors the perfect excuse to do nothing. It's China, the US, and India that needs to cut back, not us.

But if every country with "small emissions" had this attitude it would be impossible to solve climate change. As you can see in the chart, only six countries are each responsible for at least 2% of global emissions: China, the United States, India, Russia, Japan, and Indonesia.* Every other country in the world emits less than 2% and might consider themselves "negligible." But, add these countries up, and they emit 36% of the world's CO_2. That's more than China. If all these countries decided to do nothing, the biggest chunk of the world's emissions wouldn't budge. That's not an option.

To be clear: This argument is targeted at the world's richest countries, most of which fall within the "Rest of the World" box, including the UK, Germany, France, Spain, Sweden, Norway, Canada, and Australia. It's not aimed at the world's poorest countries, which have contributed very little to the problem and have tiny carbon footprints. The 142 countries that are the very smallest emitters contribute just 5% combined. So, the remaining "Rest of the World"—mostly rich—countries are still responsible for nearly a third of the total, which is plenty to make a difference.

WHAT WE NEED TO DO

As individuals, we all play a role in shaping attitudes about our country's responsibility to tackle climate change. This is especially true if you live in one of the "small emitters" where the narrative that "what we do makes no difference" has become ingrained in the culture. Push back when neighbors make these arguments. Vote out political candidates who use these claims to shirk responsibility. And lead by example by making climate-friendly choices on a personal level. The "too small to make a difference" argument

* Here I'm looking at emissions from fossil fuels and industry, so land-use change isn't included; if it was, Brazil would break into the 2% group.

taken to the extreme means that no one—anywhere—should do anything to tackle climate change.

This question is less relevant to American readers because the US *is* still a relatively big emitter. But I do still hear the argument that even the US's role is becoming irrelevant due to the rapid growth in emissions from China, India, and other emerging economies (more on this in Question 5). At a national level, countries need to focus on reducing their domestic emissions—especially the rich ones—even if they are a "small emitter" (which the US is not). But countries can also play a much bigger role on the global stage in a few ways.

- **Deploy low-carbon technologies to drive the cost down.** Low-carbon technologies need to be cheap. That's how we will make sure that every country can reduce its emissions without sacrificing human development. Countries can drive the cost down through innovation and building them at scale. There are some countries, like Germany, Spain, and China, that we can all thank for kickstarting the deployment of solar panels, making them much cheaper than they would otherwise be. Several US states—California in particular—have also played a key role developing policies in clean energy and efficiency early, before it was cost-competitive to do so.
- **Smashing glass ceilings.** No country has managed to get off fossil fuels. Or got gas and diesel cars off its roads. Or cut its greenhouse gas emissions to zero. That means everyone still has the excuse that it can't be done. But small countries can punch above their weight by resetting expectations of what is possible. They can set a model for how to transition to renewables quickly, how to build a low-carbon transport system, or how to end deforestation. Norway emits just 0.1% of the world's emissions, but it has smashed glass ceilings on the rollout of electric cars: More than 90% of new cars sold in Norway are electric.[2] The UK has become the poster child for how to wean an electricity grid off coal. When I was born it was still getting two-thirds of its electricity from coal.[3] It was still one-third just a decade ago. Last year it shut down its last coal plant.

THINGS TO BEAR IN MIND

The percentages above are based on domestic emissions—how much each country emits within its borders. But countries import goods and services, and the emissions of this "stuff" are not included in their emissions figures. If you buy a laptop that's manufactured in China, it's China that adds those emissions to its accounts, not the US.

You might be wondering how these numbers change if we account for the "offshoring" of emissions to other countries. The overall distribution of global emissions doesn't change that much.[4] Only Germany gets promoted from a small emitter—those contributing less than 2% of emissions—to a large one.

There is also a strong moral argument for why countries like the US or the UK should care, even if their emissions *today* are not a big piece of the pie. The UK emits just 0.9% of emissions, but if we add up all its historical emissions, it accounts for 4.5%. The US is the largest historical emitter by far, contributing one-quarter of the total (and there's still quite a gap to second-place China, with 15%). To be clear: This is not about blame. My great-grandparents weren't rubbing their hands thinking: "Forget Hannah's future, let's burn as much coal as we can." They needed to power their homes, and coal was the cheapest energy source they could get.

But it's still true that rich countries managed to prosper without pressures or limits on how much fossil fuel they could burn. Rich countries should get their emissions to zero quickly, not only to reduce climate change impacts for poorer countries but to "make space" in the global carbon budget for them to develop, just like my ancestors were able to.

QUESTION 5

Aren't our efforts pointless if China's emissions keep growing?

Answer: China is the world's largest emitter, but it's rolling out renewables and electric vehicles at breakneck speed.

China produces the same amount of cement in two years as the US did over the entire twentieth century.[1] In the last eight years, China has emitted more carbon dioxide than the UK has since the Industrial Revolution. It built two-thirds of the world's new coal plants in 2023.[2]

China does have a massive impact on our climate future. No question. It's the world's largest emitter, and if its emissions keep rising, we've got no chance. This can lead to a fatalism that the rest of the world's efforts don't matter.

But that excuse hinges on the assumption that China's emissions will keep increasing. Anyone who has been paying attention will know that this is far from inevitable.

China is rolling out solar and wind at a staggering rate. In a single year, it builds enough solar and wind to power the entire UK. In 2023 it installed more solar power than the US had in its entire history. It has enough solar and wind to power all its homes.[3] It produces more solar and wind power than Russia's, Canada's, or Brazil's entire electricity grid.

It's going hard on electric vehicles too. Nearly 40% of new cars sold in China in 2023 were electric. In 2024, it was getting close to 50%. From the chart you can see that it's racing ahead, leaving the United States and Europe in the dirt.

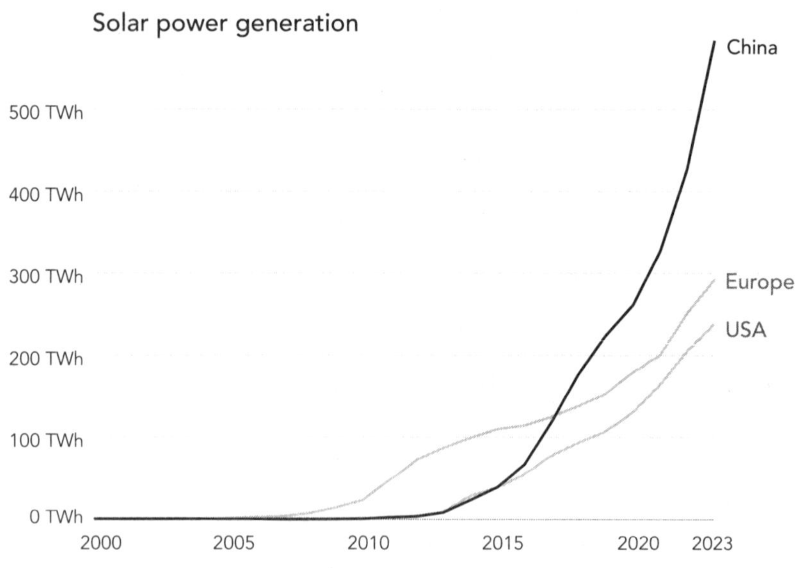

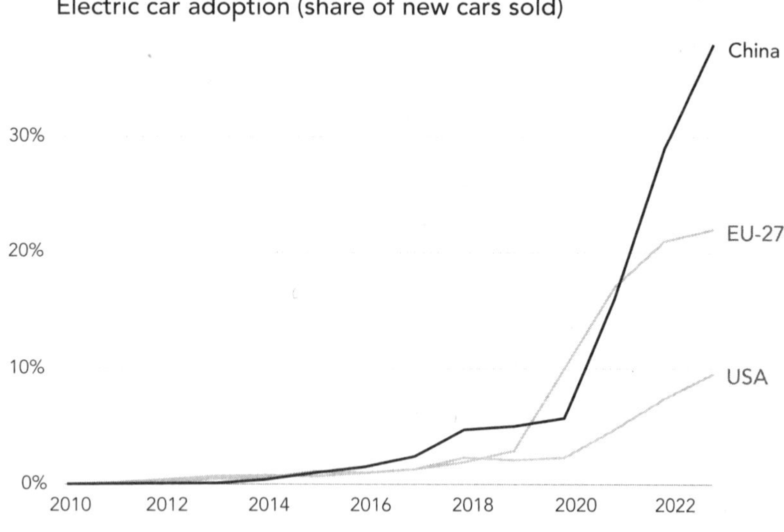

Figure 5.1
How fast are countries building clean energy solutions?
Data sources: Ember Energy and International Energy Agency.

It's not just taking the energy transition seriously at home. It's taking over the global electric car market (much to the despair of American, European, and Japanese car manufacturers). It's producing high-quality electric cars for as little as $10,000, squashing the misconception that they need to be expensive (see Question 27).

China is also dominating the supply chains of batteries, solar panels, and wind turbines. It produces around 80% of the world's solar panels, the same for lithium-ion batteries, and 40% of wind turbines.[4] The monopoly that China has on critical materials and refining capacity is one of the biggest risks to the energy transition (see Question 36). Does this sound like a country that's bowing out of the energy transition? No, China is going all in. While Europe and America drag their heels, China sees the energy transition as a major technological and economic opportunity.

China's emissions won't keep climbing. It's moving so quickly on renewables that some analysts expect its emissions from electricity to peak in the next few years.[5] Gasoline demand could soon peak too, as it electrifies its cars and transport. When China reaches peak emissions, the world will too; that's how important it is to our climate future. But that doesn't mean doom is inevitable; China is one of the few countries taking investment and manufacturing in clean energy technologies seriously. Time for others to step up and offer some competition.

WHAT WE NEED TO DO

China is moving faster than other countries on most low-carbon technologies. It's dominating the supply chains of solar panels, batteries, and electric cars. If Western car manufacturers want to stay in business, they'll have to innovate. The same is true for companies building renewable energy technologies like solar and wind. This won't just help their manufacturing industries at home; it will also drive low-carbon technologies forward, making them better and cheaper for the rest of the world.

Despite pushing on clean energy, China often gets a bad reputation in the media and public discussion. Most people are unaware of how quickly it's moving on renewables, and that makes them feel overly pessimistic about our chances of doing anything meaningful to tackle climate change. That perspective is not helpful, so it's up to us, as individuals, to counter these points whenever they come up. Make others aware of how seriously China is taking this energy transition, and people will start to demand that their own country join the clean energy race too.

THINGS TO BEAR IN MIND

What about all the coal plants that China is still building? Doesn't that mean it will keep using more and more coal for decades? China *is* still building new coal plants, but this doesn't necessarily mean that it will burn more coal. Power plants usually don't run all the time. They're turned on and off, and up and down when they're needed.

China has been using its coal plants less and less. In the early 2000s, plants were running around 70% of the time, but this has since dropped to around 50% and will keep falling.

China will use coal plants as the "balancer" in an increasingly low-carbon electricity mix. It might use coal in the way that countries like the UK or the US use gas today: to fill in the gaps when the wind isn't blowing or it's nighttime. Most of its new plants are built with technology that makes it easier to turn them on and off and ramp them up more quickly.

How much coal China burns will depend on how its electricity demand is growing, and how quickly it's building solar, wind, and nuclear. If these sources can match the new demand, coal burning will fall.

China's move away from fossil fuels will only be partly driven by climate change; what's more likely is that it will push for energy independence. Relying on others is a geopolitical liability. China is pushing for low-carbon sources that it can build and source at home.

II

FOSSIL FUELS

Fossil fuels have brought huge benefits to humans. I know this is a controversial thing to say in a book about climate change, but you and I are probably here because of fossil fuels. To take just one example, fertilizers—created from natural gas—have helped to feed around half of the global population.[1]

Even if I don't owe my life to fossil fuels, I certainly owe my *quality* of life to them. I am lucky to have been born into the safest and healthiest period of human history. The massive strides we've made in human development are not *only* because of fossil fuels, of course, but they have given me a life that my ancestors could never have imagined.

But fossil fuels are also damaging us in the form of climate change and killing millions every year from air pollution. In a world without any other energy alternatives, we might accept this trade-off. But we don't live in a world without alternatives, so we don't have to. It's the *energy* that fossil fuels bring that matters, not the fuels themselves. We mustn't let the arguments that follow get in the way of us transitioning to more sustainable forms of energy, even if fossil fuels are used as a stepping stone.

A completely fossil-free world is not going to happen overnight. Not even in the next 20 years. We should be upfront and realistic about that. It is a transition, not an on-off switch. We will still need to harness the benefits of fossil fuels where it makes sense to, while we shift to more sustainable sources.

QUESTION 6

Don't poor countries need fossil fuels to develop?

Answer: Poor countries might need some fossil fuels to develop (and that's fine), but cheap clean energy could help them leapfrog old technologies.

Countries deep in poverty use very little fossil fuels, while rich ones use a lot. The argument, then, is that poor countries need fossil fuels to develop, and it's unacceptable for us to tell them otherwise. This argument is only partly true. What people need for prosperity is *energy*, not necessarily fossil fuels. Make a graph of *energy* use per person against poverty—which is shown in the chart—and the two are tightly connected. It's just that over the last century almost all our energy has come from fossil fuels (hence why a chart of fossil fuel use against poverty would show the same). But with other cheap and reliable alternative sources, this doesn't *have* to be the future. The energy–poverty relationship will always be true. The fossil fuel–poverty one doesn't have to be.

Critics are correct, though, when they point out that no one—not least the world's richest—has the right to dictate what other countries should do.

But here's why the "poor countries need fossil fuels to develop" argument shouldn't stop us from tackling climate change.

First, fossil fuel use probably will increase a bit in the poorest countries. From a climate change perspective, this growth in the absolute poorest countries won't be a big deal. The 48 countries in sub-Saharan Africa (minus South Africa) are home to 14% of the global population. Yet they emit just 0.6% of the world's CO_2. If electricity use *tripled* in these countries, and *all* of it came from natural gas, global emissions would increase by just 0.6%.[1]

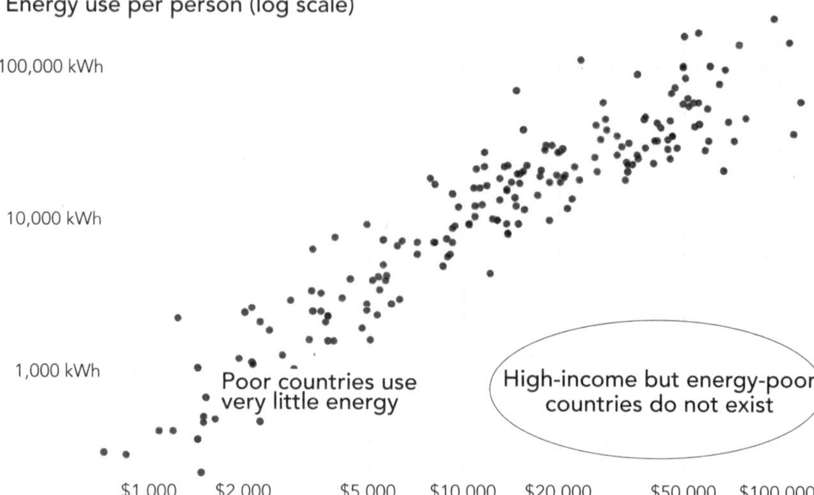

Figure 6.1

Energy use per person measured against gross domestic product (GDP) per person. One dot represents one country in 2022. GDP figures are adjusted for price differences across countries.

Data source: Energy Institute, US Energy Information Administration and World Bank.

This would have a huge impact on reducing energy poverty and improving people's lives, while the climate impact would be small.

Another area where fossil fuels could make a big difference to well-being is fuel for heating and cooking. Most people in sub-Saharan Africa still rely on wood and charcoal: This is not great for the environment but it's even worse for human health. More than 2 million people die from indoor air pollution every year, simply because they don't have access to modern fuels. Again, a temporary transition to kerosene or natural gas would save lives and probably wouldn't matter a lot for carbon emissions.

The second point is that while *some* increase in fossil fuel use among the poorest countries might be inevitable, there is also an opportunity for them to follow a much cleaner development pathway than countries like the US or the UK. Development pathways will not come with zero pollution, that is, totally "clean," but can be much cleaner.

This has already been happening. As well as China, many countries in Latin America, Asia, and Africa have been rolling out solar and wind power at a faster rate than many countries in Europe and North America (the rich "Global North").[2] Middle-income countries like Namibia, Brazil, Chile, and Morocco already get a much larger share of their electricity from solar and wind than the European Union or the United States do. And they're being deployed at the same—if not a faster—rate. The notion that solar and wind are the "green halo" technologies of the elite might have been true in the past, but in the last five years that has changed.

This focus on cleaner technologies can't come at the sacrifice of poverty alleviation and development. But it could offer cheaper and more decentralized access to energy. Households and communities in developing countries are taking matters into their own hands. Rooftop solar power has seen a boom in Pakistan as families have found a way to supply their own power. The same has happened in South Africa after the country's power providers—mostly running on coal—couldn't keep the lights on. Cheap renewables and increasingly cheap batteries are only now making this an option. In 2025, Rwanda's capital, Kigali, banned the registration of new gas motorcycles (which clock up far more miles than cars there) in favor of electric ones to tackle the city's air pollution. That puts them well ahead of most other countries in the world (including the richest ones).

Most of the world's low- to middle-income countries are in a prime position to take advantage of the clean energy revolution. They're fossil fuel importers, which means they're vulnerable to the whims and shocks of international markets. They're also rich in renewable resources—sun and wind—and are home to a lot of the critical minerals that are needed for batteries too. They don't have the same entrenched fossil fuel industries and legacy power plants that richer countries do; this gives the freedom to build a modern energy system from scratch without the sunk costs and resistance of one that's already standing. Finally, many low- to middle-income countries will take the cheapest energy technologies they can get, which usually means sourcing them from China. Geopolitical tensions between China and Western markets are driving up the price of solar panels, wind turbines, and batteries in Europe and North America. Without expensive tariffs and trade restrictions, the rest of the world will reap the benefits of super-cheap energy.

WHAT WE NEED TO DO

What can the world do to make sure that we keep making progress on reducing energy poverty, while also creating opportunities for a cleaner—and cheaper—pathway to higher standards of living?

- **Buy low-carbon technologies today.** As individuals, we can't reshape the energy choices of countries on the other side of the world. But we can help by buying and installing low-carbon energy technologies. Clean energy technologies get cheaper the more we build. By installing a solar panel, switching to an electric car, or a heat pump, you're not just reducing your carbon footprint but bringing down the cost for the rest of the world too. Most people reading this will be from a fairly rich country; it's only right that we make the transition first and make it more affordable for others.
- **Stop lecturing and dictating what poorer countries should do.** There are countless examples of "rich world hypocrisy," where developed countries tell others how to use their resources or refuse to support development if it comes with some sort of carbon footprint. During the Russia–Ukraine war, fertilizer prices rocketed. The European Union (EU) had planned to support the expansion of fertilizer production in Africa, so the continent could avoid expensive imports, but then spent a long time reconsidering that support because it would "clash with its own environmental goals."[3] Odd, considering farmers in the EU use around five to 10 times as much fertilizer per hectare as farmers in Africa.[4] It is a classic case of "do as I say, not as I do," but one that locks people in poverty and keeps families hungry. Rich countries will also often refuse to invest in *any* fossil fuel project in Africa.[5] The G7 has banned funding for "dirty investments."[6] Private European firms are frequently hit by activist lawsuits that stop them. African leaders rightly point out that these constraints are being imposed by countries that have burned huge amounts of fossil fuels for

centuries. But it's not just a legacy problem: They are being imposed by countries that are *still* doing so. It's gross hypocrisy.

- **Deliver on global finance climate commitments.** Rather than pointing the finger, rich countries could take a hard look at their own climate finance commitments and whether they're following through. In the past, they haven't. They pledged that by 2020 they would be transferring $100 billion per year to poorer countries for climate finance. But 2020 came and went, and this funding pot was almost $20 billion short.[7] This goal has finally been reached, but several years too late. They're also diverting streams of existing development aid into "green projects."[8] Climate finance was supposed to be *additional* funding, not money taken from kids waiting for life-saving vaccines or food supplies.
- **Provide better and fairer finance deals and lower interest rates for clean energy.** Solar and wind power are cheap to run, but the upfront costs can be out of reach for many. This is particularly true in poorer countries, which are considered "risky investments." If they do get a loan, it's given at very high interest rates, which make low-carbon energy unaffordable, and cripple already struggling economies with debt.[9] The cost of capital for a solar plant in middle-income countries like South Africa, India, and Mexico is around twice the cost for advanced economies (12% compared to just 5%). In low-income countries, these rates can be even higher. The world needs to finance low-carbon energy programs at much fairer prices (see Question 16 for some ideas).
- **Rich countries can support China in its clean tech revolution or compete themselves.** If rich countries want to get things moving faster, they can support China in rolling out affordable clean energy to the rest of the world *or* they can take notes, form a proper industrial strategy, and compete (see Question 5 for more on this).

QUESTION 7

Can't we just keep burning fossil fuels and capture the CO_2?

Answer: *Some* industrial carbon capture and storage will be needed, but we'll still have to burn much less fossil fuels in the first place.

A simple (at least in principle) way to stop climate change is to just capture the CO_2 that's produced from burning fossil fuels and store it underground. If it doesn't leak into the atmosphere, it doesn't cause any warming.

Industrial carbon capture and storage (CCS) is already happening in 45 facilities across the world.[1] The problem? It captures just 50 million tons of CO_2 each year. That's 0.1% of our global emissions. There could be another 50 plants operating by 2030, which would capture around 125 million tons of CO_2, bringing the total up to 0.4% of emissions. Even if we were capturing 1 billion tons per year—which is included in many "net-zero" scenarios for 2030—this would be only 2% of the world's emissions. We still need to handle the other 98%.

There's no getting around it: We are going to have to burn much less fossil fuels, and CCS is not a substitute for switching to clean energy. Yes, it's going to be hard to reach net-zero emissions without these technologies being deployed selectively. So it will probably be some part of the solution, just not a massive one.

In "Things to bear in mind" I explore what CCS could (and probably won't) be used for. Building infrastructure to capture and store CO_2 *always* makes that product—whether it's energy, cement, or steel—more expensive than it currently is. That means we need to be picky and *only* use CCS for those industries and sectors where we've run out of cheap options. The

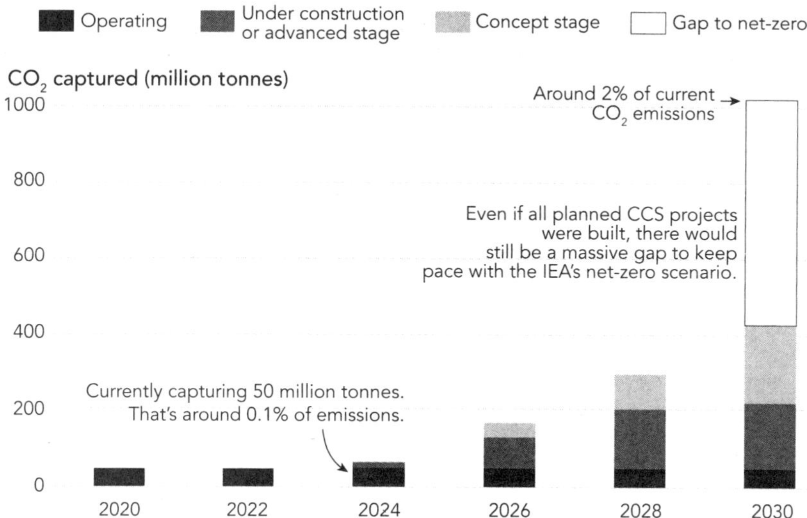

Figure 7.1

What CCS projects are capturing now, and future projections: the amount of CO_2 that's being captured by existing CCS projects, and how this might change through to 2030 with projects in the pipeline.

Data source: International Energy Agency.

prospects for CCS have shrunk over the last decade as we've found better alternatives. Fifteen years ago, the UK was talking about using CCS on coal plants. That idea is now laughable because of how cheap renewable energy has become.

Reaching net-zero emissions using lots of CCS will be far more expensive than using it selectively. Researchers at Oxford University modeled the costs of two scenarios: a high-CCS scenario, where it was responsible for cutting half of the world's emissions, and a low-CCS scenario, where it cut one-tenth of emissions.[2] The high-CCS scenario would cost more than a trillion dollars extra per year, adding up to tens of trillions by 2050.

WHAT WE NEED TO DO

Carbon capture and storage is never going to be the silver bullet that saves us. But there are ways to make sure we don't rely on it too much, without ruling it out completely.

- **Don't use the promise of CCS to delay reducing emissions today.** If we can invest in clean energy solutions that reduce or prevent emissions *now*, that's the first priority. Don't wait.
- **Innovate on solutions that don't need CCS.** CCS is the fallback solution when we can't find more cost-effective ways to handle our emissions. If we can find a cheaper way, that will reduce our need for CCS. That means innovating on zero-carbon technologies in sectors like cement and steel. We are not quite there yet but are making progress (see Part IX).
- **The world should invest *some* money into CCS so these technologies can develop, but small amounts compared to other low-carbon solutions.** Preferably money from polluters (which brings me to the next point).
- **Make fossil fuel companies invest a certain share of their profits in CCS.** This is not without controversy and would need global agreement to make it effective. But fossil fuel companies—if they were serious about their ambitions to burn coal, oil, and gas in a "responsible" way—could be capturing a lot more CO_2 than they do today. In 2022, private and state-owned companies had a net income of $4 trillion.[3] Let's say it costs $100 to capture one tonne of CO_2. If they spent even a fraction of that on CCS, they could capture billions of tonnes of CO_2 every year. Now, 2022 was much more profitable than usual because fossil fuel prices were high. In previous years, their net income was closer to $2 trillion. Still, investing some of that in CCS could have captured billions of tonnes. I'm not naive about this: There's no way we would get an agreement where these companies would invest a big chunk of their rents and profits into climate solutions. But these quick calculations show that their investments could be much higher.

THINGS TO BEAR IN MIND

CCS is a technology that has failed to live up to its promises. By 2020, 149 projects were supposed to be online and capturing carbon, but more than 100 of them were canceled.[4] Eighty percent of projects in the United States have crumbled because they were not financially viable, or the technology wasn't ready to roll out.

What could CCS be used for? Not transport, because those emissions are fragmented: They come out of billions of cars, trucks, and planes, not a single power plant that you can put a capture technology on. That means it's not a solution for most of the world's oil; it can only be used in oil refineries, which are a small fraction of the oil industry's emissions.

It probably won't help with the bulk of our electricity needs: we have cheap solar and wind, supported by hydropower, nuclear, and maybe emerging technologies like geothermal. It *could* play some role in making up the final part of an electricity mix that's almost completely powered by renewables. Solar and wind can provide the first 70–80% of electricity very cheaply. The problem for some countries—particularly those at higher latitudes—is getting the final 20–30% of stability because wind and sun are not available all the time. Gas with CCS is one of several technologies in the running to be a small part of the final portfolio to fill in the gaps.

Industries such as cement, steel, and fertilizers might also have to rely on CCS. These make up a surprisingly large share of the world's emissions: Cement and steel alone are close to 15%. CCS could be their only or most cost-effective option, although exciting innovations are coming through that could be better and cheaper. The cheapest option will win, and I expect that in some of these sectors CCS will eventually drop out of the race.

QUESTION 8

Can we transition to clean energy fast enough?

Answer: Historical energy transitions were slow, but clean energy is growing faster than any energy source in history.

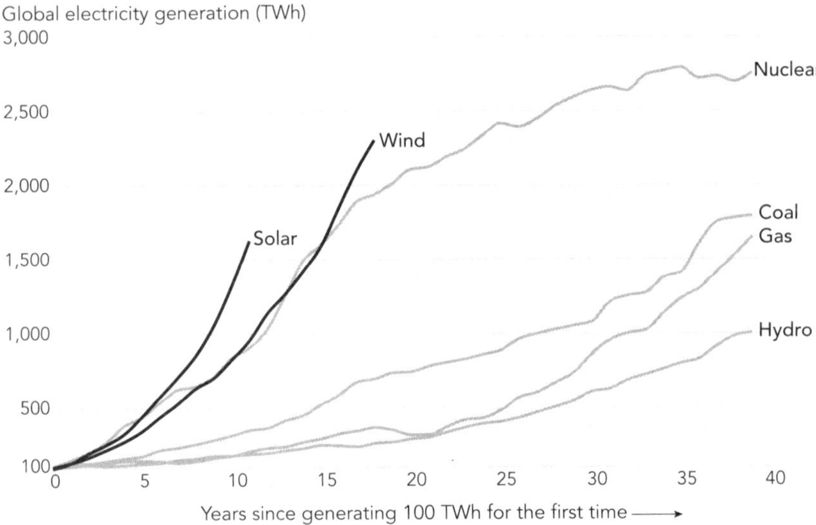

Figure 8.1

How quickly energy sources have been rolled out over time: The amount of global electricity generated after a source reached its first 100 terawatt-hours of production. This is roughly the point where sources start to scale. Energy sources reached this scale at very different points in history, but they've all been normalized to the year 0. It's shown as the years since *each source started to scale*.

Data source: Ember Energy and Pinto et al., 2023. Based on a chart by Nat Bullard.

The world needs to move away from fossil fuels quickly. Not within a few years as some activists call for (that's not possible), but within decades. If you look at the history books, though, you'd say this was impossible.

Parts of the world have gone through several major energy transitions. Wood to coal. Coal to oil. Coal and oil to gas. It took at least 60 to 70 years for these shifts to happen. And since many people in the world still rely on wood for fuel, you could easily argue that, globally, we're still in the middle of these transitions—centuries after they started. These timelines are much longer than the 30 years we have to move the world to clean energy now.[1] But here's why historical energy transitions might not be a good lens for seeing into the future:

1. Clean energy is already growing much faster than fossil fuels did in the past. In the chart, you can see that it's growing several times faster than any energy source in history. A decade ago, we'd have said that this growth wasn't possible. We're already in unprecedented territory.
2. Energy transitions haven't *always* been slow. They have been when we look at *global* changes, but these hide the fast transitions that have happened in some countries. The UK's move from gas to electric light happened within two to three decades in the early 1900s. And again a century later: When I was born in the 1990s, almost two-thirds of the UK's power was coming from coal; now we're coal-free. The transition from horses to cars in the United States similarly took just 30 to 40 years in the early twentieth century. In response to the 1973 oil crisis, France built out a huge nuclear energy fleet within two decades. These examples dismantle the notion that energy transitions are always slow.
3. Solar and wind are modular technologies, so they can be built quickly. You can manufacture the same solar panels over and over. Fossil fuels tend to require larger centralized projects that vary a lot from place to place. Building a coal or gas plant takes twice as long as installing the same amount of renewable energy. Our energy is changing from being a commodity to a technology; something more akin to computer chips or microprocessors. In the digitized world we live in today, knowledge about solar and wind can spread quickly between countries. This wasn't

the case when fossil fuels were breaking through in the nineteenth and twentieth centuries.

4. Renewables can also piggyback off existing infrastructure. Solar, wind, and batteries can plug into an existing grid. This couldn't happen with the transition from horses to cars, or coal to gas. Countries had to build roads and highway systems and build out vast liquid pipelines.

Finally: there is the "minor" detail of climate change. It forces us to move faster than we've ever done before.

WHAT WE NEED TO DO

- **Combine the power of markets with guidance from policy.** If we were to let these technologies and markets roll along, the transition *would* happen, just not fast enough. Climate policies and targets give it direction and extra momentum to get us there quicker. There are plenty of examples of this. The German government set out a bold strategy to support renewable energy in the 1990s when it was still very expensive. The Chinese government has implemented an ambitious industrial strategy for clean energy, providing large subsidies and support for low-carbon technologies. More recently, the US government got involved. Its Inflation Reduction Act provided policy support and incentives—potentially up to hundreds of billions of dollars—to a variety of clean energy projects to get them moving faster. How this now plays out with a new government remains to be seen.
- **Build out and modernize our electricity grids.** To get electricity into our homes, offices and manufacturing plants, we don't only need the energy sources themselves, but a well-connected grid to move that power around. We need the pylons and the cables to transport electricity from the source to where we need it, and that whole system needs to be perfectly balanced so we always have enough power, but not too much. Renewables could be growing a lot faster if our grid systems were ready to connect them. Unfortunately, the electricity grids—mostly the cables and transformers—in many countries are old and have been poorly

maintained. This makes it even more difficult to add more power plants to the system.

- **Reform planning permissions for clean energy projects.** Seriously, just get out of the way. If you want people to build stuff, you need to make it much easier for them to do so—5,000-page application forms are not going to cut it. We need to build responsibly and thoughtfully, but if we want to build at all, we're going to need to cut some of the red tape.
- **Learn from international experiences.** This is a *global* transition, not a national one. We can learn from each other about how to do it efficiently, and how to overcome barriers.
- **Stop underestimating the speed of emerging technologies.** Solar, wind, and batteries have grown far faster than most people would have imagined just a decade or two ago. Many energy analysts and commentators (including me) underestimated the takeoff of solar and wind. You might have seen the now infamous charts comparing forecasts from organizations like the International Energy Agency to reality.[2] Solar growth has torn past forecasts year after year, and not by a small margin. The Intergovernmental Panel on Climate Change scenarios have also vastly underestimated the rapid uptake of solar.[3] The frameworks we've used to model energy systems in the past are probably not fit for our energy future.

QUESTION 9

Are we even transitioning to clean energy if we keep adding more fossil fuels?

Answer: The energy transition is already happening in many countries, and the world is on the brink of a global transition.

Some people claim that there's no "energy transition" at all. They point you to a chart showing that, globally, both fossil fuels and renewables are rising. Renewables are not replacing fossil fuels; they've just been dropped on top of an ever-increasing demand for coal, oil, and gas.

But that's an oversimplification. It lumps every country—those at very different stages of development—together. And it fails to recognize the important role that renewables have, and currently play, in keeping carbon emissions at bay.

There are three roles that renewables *could* play in our energy system:

1. **Energy addition:** Fossil fuels grow as normal, and renewables are simply added on top.
2. **Energy displacement:** Renewables are filling a hole in demand that would otherwise be filled by fossil fuels.
3. **Energy transition:** Renewables are replacing fossil fuels.

I've mapped out each of these scenarios in the chart. The first—where renewables are not impacting fossil fuels at all—would be disappointing. But I can't find any examples where this is happening.

Many countries—and the world as a whole—are in stage two. Energy demand is increasing because populations are growing, more people are getting access to energy, and they're being lifted out of poverty. That's a great

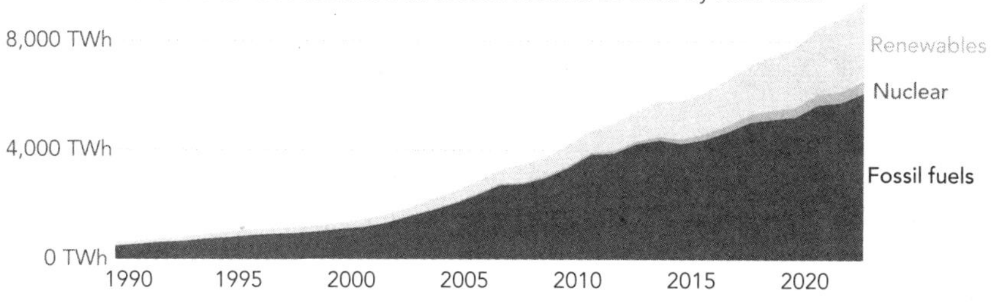

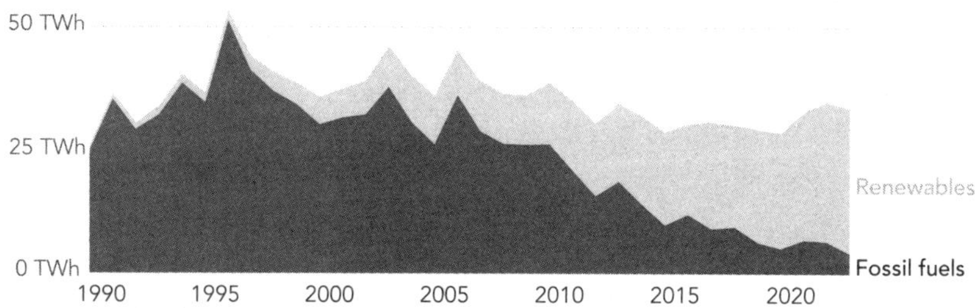

Figure 9.1

Some countries are replacing fossil fuels while others are displacing them: electricity production by source in a scenario of energy displacement in China, and energy transition in Denmark. There are no obvious examples of countries where renewables are simply adding to demand. *Data source*: Ember Energy, 2024.

thing. Renewables are expanding quickly and meeting a lot—but not all—of that demand. Fossil fuels are also increasing, but far less than they would be without renewables.

Most rich countries (and some on lower incomes) are in stage three. Their energy system is transitioning; renewables *are* replacing fossils. More than half of the world's countries have passed "peak fossil fuels" in the electricity sector.[1] More than half of those countries did so *while electricity demand was rising.*

The United Kingdom, Denmark, and Finland are obvious examples. Since peaking in 2008, the amount of fossil fuels being burned for electricity in the UK has fallen by almost half. In Denmark by almost 80%, and in Finland by more than two-thirds, since 2000. The US is also part of this group, but progress has been slower with fossil fuels dropping from 72% to 59% of the electricity mix (although grouping "fossil fuels" together misses the fact that it has also had a massive transition from coal to gas, which is still a positive for the climate).

Although globally we're still in stage two, in the next few years we're going to enter stage three. In 2022, renewables met 92% of the extra electricity demand. The other 8% came from fossil fuels; a tiny amount compared to the past. The growth in renewables is soon going to outpace demand growth. Fossil fuels are then going to tumble. Leading energy analysts now expect that coal, oil, and gas will all individually reach their peak globally in the next five years.

The world *will* transition away from fossil fuels. How quickly will depend on us, and whether we let many of the misconceptions in this book stand in the way of progress.

WHAT WE NEED TO DO

How do we make sure we're driving the energy transition, and not just energy displacement?

- **Deploy clean electricity sources as quickly as we can.** This is obvious. Solar, wind, hydropower, nuclear, geothermal: We need to be building as much as we can as quickly as we can. *How quickly* depends on the country. The US, for example, has set a target to fully transition away to low-carbon electricity by 2035. The UK has a more ambitious timeline of 2030. To get there it will need to build offshore wind and solar about three times faster than it has in the last two decades. At a global level, analysts expect clean electricity sources to cover all the growth in electricity demand in the next three to four years. That means we'd be displacing *all* of the additional coal and gas power, but it wouldn't be enough to start biting into existing fossil fuels. To do that, and really tip in to a transition, we'll need to double or triple the pace of deployment by 2030 and accelerate even further from there.

- **Keep existing nuclear plants running.** Some countries have shut down perfectly good nuclear power plants ahead of schedule. This is one of humanity's biggest own goals. Rather than replacing fossil fuels, renewables are replacing nuclear, another low-carbon electricity source. Germany and Japan, I'm looking at you.
- **Electrify as much as we can.** I've mostly talked about the transition in the *electricity* sector, which is only part of our energy demand. We need to do the same for transport, heating, and industry. That means we need two transitions to happen at once: We need to make our electricity sector clean *and* electrify other industries as much as we can.

Beyond electricity production, the transition looks less impressive because this second step—electrification of other industries—is still in the very early stages. Electric cars and heat pumps are only just taking off. But that transition will come. Sales of new gas and diesel cars already peaked in 2017.[2] We need to get fossil fuels out of our electricity grids as soon as possible because, in a decarbonized world, electricity will dominate our energy systems. But we need to make sure that the transition to electric cars, heat and industry is moving quickly too. For more on that, read Parts VI, VII and IX.

THINGS TO BEAR IN MIND

Let's look at an example. Between 2010 and 2022, global electricity generation increased by 7,590 terawatt-hours (TWh). Renewables supplied more than half—57%—of that increase. The rest came from fossil fuels. In a world without renewables, fossil fuels would have made up almost all that new demand. If we *hadn't* deployed more renewables, the world would be emitting between 2.1 and 3.6 billion tonnes (1 metric ton or 1 tonne = 2,200 lbs) more carbon dioxide every year.* That's more than the emissions of the European Union or India.

* I've calculated this by assuming that all the electricity from renewables was produced from coal or gas. If it was replaced by just coal, an extra 3.6 billion tonnes would be emitted each year. If it was gas, this would be 2.1 billion.

QUESTION 10

Will we even be able to produce enough clean energy to replace fossil fuels?

Answer: Most energy that comes from fossil fuels is wasted; by moving to clean energy we'll need to produce far less.

The world burns massive amounts of fossil fuels every year. That's why we're in this climate mess. But the challenge to produce the same amount of energy from renewables or nuclear seems overwhelming.

The good news is that we don't have to. This transition is not a simple swap, with one unit of solar energy replacing one unit of coal. In reality, one unit of solar can replace two or three units of coal. That's because our current energy system—largely based on burning stuff—is woefully inefficient. Most people are totally unaware of how much energy the world is wasting. We're burning fuel and it's not doing anything to improve our lives.

Moving to a low-carbon economy comes with massive gains in efficiency in two key stages:

1. **Replacing inefficient fossil fuels in the electricity mix.** When we burn coal in a power plant, only around one-third of its energy is converted into electricity. The rest is wasted as heat. Natural gas is slightly better, but half is still wasted. When we move to renewables, there is hardly any waste. Let's say we need to produce 100 TWh of electricity for a city. To get that from coal, we'd need to burn the equivalent of 300 TWh (since two-thirds would be wasted). From gas, we'd need 200 TWh. But from solar and wind, we'd just produce the 100 TWh. Nothing is being burned and going to waste.

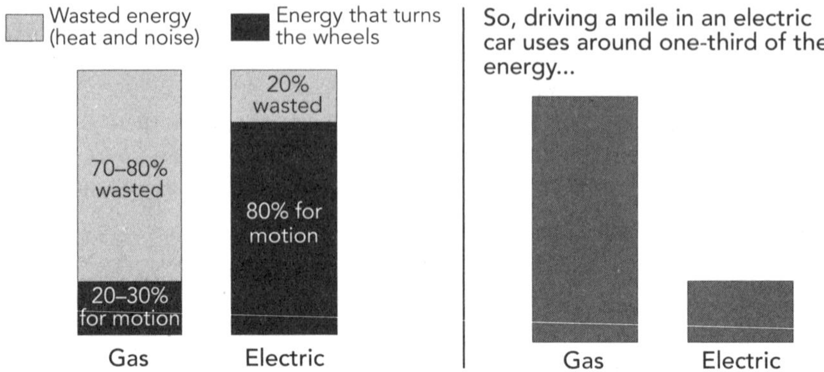

Figure 10.1

The energy use of electric versus gas cars: Electric cars are much more efficient than internal combustion engines, using three to four times less energy. They lose much less heat in the engine, and regenerative braking means they can use energy losses from braking.

2. **Taking advantage of the efficiency of electrification.** One reason we want to run much more of the world—our cars, heating, and industry—on electricity is it makes decarbonization much easier. We can't make gas low carbon, but we can run our cars on solar and wind if we power them with electricity. But another reason is that electrification comes with huge efficiency benefits that we don't talk about enough.

Take the example of driving a car. A gas car is only 20% efficient.[1] For every dollar of gas you put into a car, only 20 cents is used to move the vehicle. The other 80 cents is wasted, mostly as heat from the engine. What a rip-off. In an electric car, it's the opposite: Around 80% of the energy going in is used to turn the wheels, and just 20% is wasted. That's because it hasn't got an inefficient combustion engine, and it has regenerative braking so the battery recharges as you stop and start. This means that driving one mile in an electric car uses a quarter to a third of the energy of driving a mile in a gas one. There are more Questions and Answers on electric cars in Part V.

Electric heat pumps are more efficient than gas boilers, and the same is true for electrified industrial processes. Going electric is one of the most energy-efficient switches we can make.

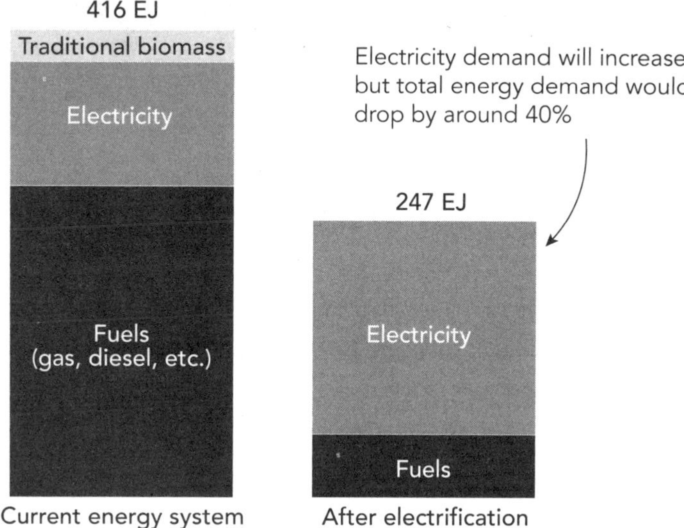

Figure 10.2

The amount of energy the world uses today, and how much it would use after electrification: Electrification will make our energy systems much more efficient.

Data source: Nick Eyre, "From Using Heat to Using Work: Reconceptualising the Zero-Carbon Energy Transition," 2021.

Let's step back and think about what this means: We don't need to replace *all* our existing energy (most of which comes from fossil fuels) with clean energy; we'd get the same societal benefits from much smaller amounts of energy. Take the previous car example: If the world's vehicles went electric overnight, energy use for road transport would suddenly drop by two-thirds to three-quarters. That's without any change in how far people drive, or the size of their cars.

If the world were to electrify all the sectors that *could* be electrified, energy demand would drop by around 40%.[2] Combine this with the efficiency gains of moving away from burning fossil fuels for electricity, and the drop would be even greater.

This means that energy use in rich countries could go *down* over the next few decades. In poorer countries, energy demand will still increase because people are moving out of poverty and achieving higher standards of living. But moving to a low-carbon economy means they will be able to do so in a much more energy-efficient way. That's all thanks to a shift away from "burning stuff": the way humans have made energy for over a million years.

WHAT WE NEED TO DO

These recommendations will make the challenge of replacing fossil fuels with clean energy as easy as possible.

- **Electrify everything.** Cars, motorcycles, trucks, heaters, steel plants, ferries, short-haul flights: If it can be electrified, then do it. Of course, as individuals, we play a crucial role in this: If you can afford to, swap the gas car, gas boiler, and gas cooking stove for their electric cousins. It will be good not only for the climate but for your bank balance and health in the long run too.
- **Stop burning stuff.** For most of human history, we just accepted that a huge amount of wasted energy was inevitable. But that's not the case anymore. For the first time, we can produce energy for 8 billion people without burning stuff. What an opportunity. We're lucky enough not only to have a front-row seat but to be a player on the field, making it happen.
- **Spread the word about the efficiency benefits of moving to clean energy.** People would be shocked if they knew how inefficient our current energy system is. Let them know.
- **Don't panic when electricity demand goes up.** Remember that electricity demand *is* going to increase as we electrify more parts of our energy system. Crucially, *total energy* use should fall, even though we're using more electricity. In some sense, rising electricity demand is a sign of success; don't panic when it does happen.
- **Combine with other efficiency measures.** Of course, we shouldn't forget about the other things—which don't involve electrification or moving to renewables—we can do to make our energy systems more efficient. Again, these are things we can do as consumers: insulated homes, efficient air conditioners and central heading, and fewer SUVs on the road would make this energy transition even easier. Let's do them all.

QUESTION 11

Won't we need fossil fuels to build low-carbon energy in the first place?

Answer: Solar panels and wind turbines generate far more energy than is used to build them, and pay back fast (often just a few months).

Some people say that a move to renewables can't be done because we'll need too many fossil fuels to build the infrastructure. But they seem to be oblivious to what the word "transition" means. This is how all our energy transitions have worked in the past. You use your existing energy source to build the infrastructure for the new one.

So, the simple fact that we'll need to use existing energy systems (mostly fossil fuels) to build new ones (renewables and nuclear) is a weak argument against the transition.

If we look at energy sources and technologies using a metric called the "energy return on investment" (EROI), this tells us how much energy we will produce for every unit of energy we use to build that source. So, if a solar panel produced 15 units of electricity for every unit of energy needed to build it, it would have an EROI of 15.

Having an EROI of less than 1 is bad: You actually *lose* energy. You'd put 1 unit of energy in but get less in return. Having an EROI of 5 or less isn't great either. Sure, you produce more energy than you put in, but it's still an incredibly inefficient system. When we properly account for energy waste, renewables and nuclear power tend to have far higher EROI values than the "threshold" of 10 that we'd want to achieve.[1]

Another way to think about this is to consider how long it takes for a solar panel or wind turbine to "pay back" the energy that was used to produce it. For solar, it's now less than one year.[2] In the first year, it's mostly paying its debt on the energy it took to produce it, but afterward it's giving a net gain in energy back to society. Solar panels tend to last 25 to 30 years, so that's a very good deal: It produces 25 times as much energy as it consumes. For wind or nuclear, the payback period is just months.

Yes, we'll initially need to use some fossil fuels to build our low-carbon energy system. But it won't be in huge quantities. It won't eat up all our supplies. And it won't go on forever. It's a *transition*.

WHAT WE NEED TO DO

There's not much that you or I can do about the energy density of different energy sources, but there are a few things that can make a difference.

- **Accelerate the energy transition so more manufacturing energy is coming from renewables.** In supply chains, companies can accelerate the transition to make sure that less of the energy needed to manufacture solar panels and wind turbines is coming from fossil fuels. The innovators and managers working on these strategies have a crucial role to play. But as consumers, we do too. Putting pressure on companies to clean up their act and decarbonize gives them even more impetus to do so.
- **Invest in clean energy projects for mining.** Mining companies are starting to invest in renewable energy projects to power their processes. Manufacturers are trialing electrified drills and heavy machinery to reduce their dependence on oil and gas. Companies that produce the final products are investing in processes that make them more energy-efficient, which not only reduces their environmental impact but improves their bottom line too.

THINGS TO BEAR IN MIND

A nerdy point on EROI, for those interested.

EROI is nonlinear. It's often described as having an "energy cliff." At very low values of EROI, differences matter a lot. Going from one to two is massive. There, we're on the very steep side of the cliff. Two to three is a bit less steep as we start to climb it. A technology that has an EROI of 10 will deliver 90% of its energy as net gains to society, and after that the gains will be much less noticeable. So, we should fret less about whether our EROI is 10, 20, or 30 and focus on whether it matches the "minimum" EROI that society needs. Ask energy boffins what the "minimum" EROI we should accept is, and most say something in the range of 5 to 10.[61]

QUESTION 12

Won't we need more fossil fuels to keep up with artificial intelligence?

Answer: Artificial intelligence will increase electricity demand, but most of this could be met with clean energy if we're smart about it.

Artificial intelligence (AI) has become synonymous with "energy guzzler." A search with a large language model (LLM)—like ChatGPT, Claude, or Gemini—is thought to use 10 times as much energy as a Google Search.[1] I say "thought to" here because this statement came from the Alphabet chairman in 2023; since then, tech giants have not been open about how much energy AI uses. It would be good if there was greater transparency for these companies on actual, measured energy use (more on this in "What we need to do"). So, does the birth of artificial intelligence means the death of our climate goals? I don't think so. At least, it's not guaranteed. While AI will increase electricity demand, it probably won't be on a scale that clean energy can't rise to.

Data centers currently use only a few percent of the world's electricity (estimates vary from around 1 to 3%).[2] The big question, though, is whether this will explode with the rise of AI. Probably not. At least not in the short to medium term. In 2024, the International Energy Agency (IEA) looked at where the growth in global electricity demand was going to come from by 2030.[3] In figure 12.1, you can see the results.

It projects that annual demand could increase by 6,000 TWh by 2030. Data centers represented just 3% of this growth. Other things that we'll cover in this book—like electric cars, heating, air conditioning, and industry—accounted for a lot more. So, yes, data centers will add to that demand, but they won't dominate.

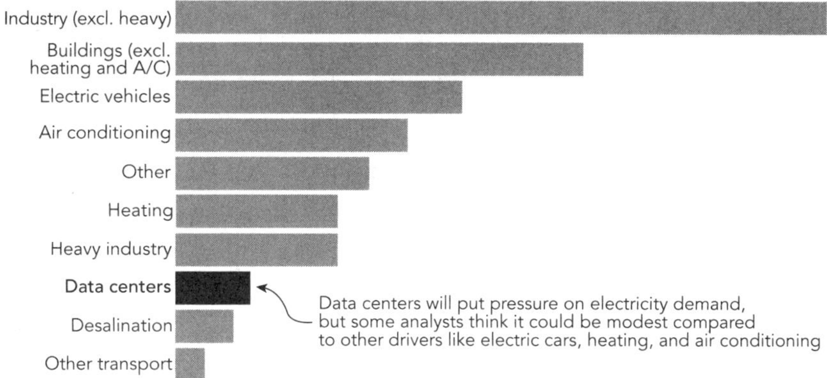

Figure 12.1

What will drive the growth in global electricity demand between 2023 and 2030? Data center demand is uncertain, but some leading analysts think it will be modest compared to other drivers of electricity demand.

Data source: International Energy Agency, 2024.

The IEA could be underestimating the growth of AI. Other analysts think that AI's electricity demand could double or even triple by the end of the decade.[4] Now, this doesn't mean that total electricity demand would triple—just demand for data centers. In other words, you could see their share of global electricity increase to around 6%. These scenarios are certainly plausible. And we would need to up our game to meet this demand from clean energy. But this is quite different from wild claims about AI overrunning our energy systems. The biggest tension for AI and climate change is not the total amount of energy that's required, but where this energy is needed and how quickly the supply will have to be built.

Energy demand for data centers is very localized.[5] While data centers use just a few percent of the world's electricity, in some countries they use a lot more. In Ireland, it's 17%, and more than five US states used at least 10% of their electricity for data centers. Training and scaling AI, then, will put increasing pressure on local electricity grids. They will need constant power, and they need this additional power now. In fact, given the AI arms race—with companies advancing AI rapidly to beat the competition—they really needed it yesterday.

This is why there's a concerning grab for more natural gas, which can run 24/7 and often offers existing infrastructure that can be tapped into. But while there might be a short-term rush for gas, in the longer term we'll need to get most of the additional electricity from clean sources if we want to tackle climate change. What are our options?

The cheapest option in many places will be the combination of renewable energy and batteries. The issue, though, is that there will still be some periods where solar or wind plus batteries won't be enough. That's fine: Combine it with other, more stable sources of clean energy like hydropower, nuclear, or geothermal. In fact, even burning small amounts of gas to cover those rare occasions could be fine, and much more climate-friendly than running them on gas all the time.

Nuclear power is a growing favorite with the big tech companies. Amazon, Microsoft, and Google have all announced large investments in advanced nuclear technologies. This makes perfect sense if you want 24/7 clean power. The problem is that building nuclear plants takes time (especially in the US and Europe—see Question 22), so while this could be a solution in the long term, it's not going to meet demand tomorrow. If these companies want nuclear power sooner, they could focus on reopening closed nuclear plants; that's what Microsoft is trying by making a deal to reopen Three Mile Island.[6]

Meeting rising demand for AI through clean energy should be possible. But it will need innovative and bold leadership not to fall into the trap of meeting all of this demand with the easy option: natural gas.

Finally, fears around the energy demand for AI are only one side of the coin. Flip it over, and we have its potential to make fighting climate change a lot easier. It could help us find hidden mineral deposits and dig them out more responsibly; it could help us optimize and balance our electricity grids so that we have just the right amount of solar, wind, nuclear, and other energy sources; it could forecast weather faster and at higher resolution than humans, helping us find the perfect spots for solar and wind farms and protecting communities from the damages of climate change.

Yes, the gold rush for energy for AI seems scary and overwhelming. We need to find a way to manage it responsibly. But to shut the door on it might make fixing climate change harder, not easier. Regardless of whether we'd want it, the genie is already out of the bottle and it's not going back in.

WHAT WE NEED TO DO

None of us know what the future of artificial intelligence looks like, but there are things we can do to take the opportunities it offers without jeopardizing our climate goals.

- **Don't give up on AI.** It's tempting to see headlines about how energy-guzzling AI is and vow never to use it. That would be a mistake. Most people can use ChatGPT, Gemini, or Claude quite a lot, without much of an impact on their carbon footprint. Asking ChatGPT 100 questions uses the same amount of energy as streaming a 20-minute video. Or using an electric heater for one minute. Your personal use of text-based AI is small compared to the "big stuff" like driving, flying, heating or cooling your home, and what you eat.
- **Invest in the energy efficiency of training and querying AI models.** This is not the first time people have panicked about an unsustainable explosion in energy demand from the internet. The late 1990s and early 2000s were filled with scary predictions that data centers would guzzle up more than half of America's electricity. While demand for the internet did skyrocket, energy use didn't. Between 2010 and 2018, global data center computing increased by more than 550%. Yet energy use in data centers increased by just 6%.[7] This is because of "Koomey's Law," which describes the huge gains in the efficiency of computing. Now, there's no guarantee that this will continue—in fact, I don't think more efficient chips and processes will keep up, hence why we'll need some extra energy. But to keep this demand in check, investments in efficiency will be crucial. This is already happening because efficiency will be the edge that tech companies fight for.
- **Stop hyping unknown future scenarios.** We're going to need more energy to power AI. This doesn't mean global energy demand is going to double or triple. Severely underestimating how much energy we'll need is a

risk because we'll be unprepared. But vastly overestimating how much we'll need is also damaging: We risk building a flurry of new gas plants to power data centers that will never arrive. Hyping AI energy demand provides the perfect cover for companies that want to see a resurgence of fossil fuels. Concern is good; panic is not. It leads to rushed decisions that make things worse.

- **Demand transparency from tech giants.** Tech companies—with valid concerns about competition—don't publish how much energy it takes to train and operate their AI systems, so researchers need to find other ways to come up with a best estimate (in "Things to bear in mind," I explain a few ways they do this). Tech CEOs should stop going on podcasts talking about "building AI responsibly" if they're not going to be transparent. That's just PR.
- **Keep pressure on tech companies to meet their climate targets.** Most tech companies—from Meta to Microsoft and from Amazon to Google—have set climate targets to get their emissions to net zero. AI is not an excuse to roll back these promises, but to push harder and faster for low-carbon energy instead. Many of them are taking this seriously and exploring different ways of getting clean, reliable power. This could end up being good for the world if it drives innovation that the rest of us benefit from too.

THINGS TO BEAR IN MIND

I want tech companies to be transparent, so I should be too. The first few drafts of this book didn't have a question on artificial intelligence. This is a book grounded in data, and the truth is that there was very little data on the energy and climate impacts of AI. I didn't want to just hand-wave.

But as the months passed, it became obvious that I had to include something about it. People constantly ask about it at talks, in interviews, and online. Leaving it out would be a colossal elephant in the room.

In the last year, more numbers and projections have appeared, which means I can give some summary of the best research out there. But these technologies are moving quickly, so I'd take everyone's projections with a pinch of salt. You'll also notice that I've only focused on medium-term projections to 2030. Anyone trying to predict the development of AI any further into the future is simply grasping at straws.

You might be wondering how researchers estimate the energy demand of data centers and AI if companies rarely make their figures public. There's a variety of ways to do it. One basic way to get a sense of the scale of AI energy demand for *users* (us *using* services like ChatGPT or Gemini, rather than training them) is to calculate how much extra energy Google would need if every one of its searches were done with an LLM. As it turns out, not very much.[64] It's also worth pointing out that Google wouldn't be able to make that transition very quickly because the *hardware*—the servers and the chips—would be a bottleneck. The world couldn't build them fast enough.

Another way to get a sense of scale is to look at how many servers and chips NVIDIA—the company that has almost single-handedly supplied the world's hardware for AI—has sold in the last few years. Researchers can then estimate how much energy these servers would use if they were all running at full capacity. Again, it's not a lot. To get some projection of future demand, you could guesstimate the maximum number of chips that NVIDIA and other emerging companies would be able to manufacture over the next few years and calculate how much energy they would consume.

These are just a few simple examples of ways that researchers can get a grasp on the scale of AI energy demand. In reality, they use more complex methods to back these numbers up. But this still involves taking small nuggets of public information and building a model of the digital world from it. Not ideal, but the best they can do with what's available.

QUESTION 13

Won't a lot of energy workers lose their jobs?

Answer: Jobs in fossil fuels will decline, but clean energy is creating even more of them.

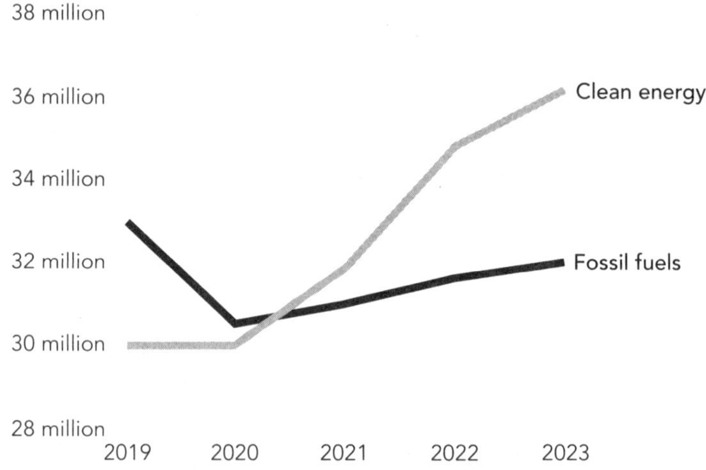

Figure 13.1
Global number of workers in clean energy and fossil fuels: The total number of energy workers globally is rising.
Data source: International Energy Agency, 2024.

People worry a move away from fossil fuels will lead to massive job losses for the energy sector. The opposite is true: There will be *more* people working in energy in the decades ahead, not fewer. As you can see in the chart, the number of people working in clean energy overtook fossil fuels in 2021, as far more money is invested in it.[1]

The *total* number of people working in any energy job (fossil fuels and clean energy) will increase. It's already rising: The number of these workers increased from 63 to 68 million between 2019 and 2023. By 2030 this could reach 72 million. If the world was to push harder on low-carbon energy, it would be an extra 10 million by 2030.

The IEA chart looks at the number of jobs *globally*. But is it also true for richer countries, where there's a concern that jobs get displaced to other parts of the world? The answer is yes: Energy-related jobs will also increase across Europe and North America. Clean energy is generating a manufacturing boom in the United States: Energy-related jobs grew by 3.8% in 2022, which was faster than the growth for the economy as a whole.[2] Clean energy grew at 3.9%. In the UK, "green jobs" are growing at four times the market rate.[3]

While some jobs—such as mining for particular minerals or materials—are location-specific, most jobs related to clean energy cannot be "offshored" to other countries. Workers are needed *in the country* to install solar panels and wind turbines, develop electric vehicle charging infrastructure, install heat pumps and insulation in people's homes, and build the grid networks that we need to keep our power running. Those jobs can't be done on another continent.

Thirty-two million people globally work in the fossil fuel industry today. That's around 0.4% of the global population. Climate change affects 8 billion of us today, plus the hundreds of billions that could come after us. We can't shove that 0.4% aside; we should help them to retrain and find new jobs.

New and exciting industries are emerging, which could deliver more skilled, well-paid jobs. Not to mention jobs that lead to a cleaner, safer future. This is important for public support: Studies have found that job creation in low-carbon energy improves support for wider climate action and decarbonization.[4]

WHAT WE NEED TO DO

If we don't take this problem seriously, we will not only leave whole communities behind but will also lack the skills we need to make the energy transition a reality. Here are some things we can do to avoid this.

- **Invest in vocational training for clean energy and manufacturing.** A bigger problem in many countries is not that jobs are disappearing, but that there aren't enough skilled workers to fill the new ones. Many countries have underinvested in their electricity grids, so there aren't enough trained workers to build them up again. Engineers are trained to install gas boilers, not heat pumps. We need an urgent upskilling of people to work in the new energy sector. That means fewer university graduates and academics (like me) and more workers with practical skills. Governments need to invest in training so that this doesn't slow us down.
- **Support local communities that have relied on fossil fuel industries.** Energy and industrial transitions in the past have often left communities behind. This is still happening: I grew up near Scotland's largest oil refinery; it's set to close this year with hundreds of workers losing their jobs. If we fail to provide support for areas that are dependent on fossil fuel industries, we risk widening economic inequalities and creating a backlash against climate action. Governments and societies need to support these communities to make sure there are new and existing opportunities available so they can transition to a post–fossil fuel era.
- **Retrain existing fossil fuel workers to transition to other well-paid jobs.** While new jobs will be created, we shouldn't forget or deny that millions of fossil fuel workers *will* lose their jobs. We need to support workers in these industries—who have developed invaluable skills and experience—to find good work in other places. Many of those in existing fossil fuel jobs do have transferable skills.
- **Do not put tariffs on low-carbon technologies.** One argument for slapping tariffs on cheaper solar panels and batteries (mostly from China) is that it makes *domestic* manufacturing in the US or Europe more cost-competitive, and therefore boosts blue-collar work at home. This misses the fact that most clean energy jobs are *not* in manufacturing; they're in the deployment, installation, operations and repair of solar and wind farms, grid networks, and storage. In the US, over 80% of jobs in solar power are not in manufacturing. The same is true in Europe. By making solar panels more expensive, people deploy them more slowly and this

reduces the total number of jobs for the energy industry. Tariffs hurt rather than help domestic jobs.

THINGS TO BEAR IN MIND

What does the transferability of energy jobs look like?

Around 6 million workers could lose their jobs in the traditional car industry (gas and diesel ones) by 2030. But almost two-thirds of those workers could be transferred to other industries (such as electric car manufacturing) with retraining. Car manufacturers are already investing in retraining because they need skilled workers to compete in a sector that's going electric. Volkswagen recently pledged to train 22,000 to work on electric vehicles at one of its European plants. Mercedes-Benz has invested more than €1.3 billion in retraining, and Ford in the US has invested over $11 billion.

There could be a 2.5 million job loss in the oil and gas sector, but three-quarters of those workers could be moved to other parts of the energy industry. Of course, that will rely on investment in retraining, and support for the remaining workers that can't be redeployed. My own country's track record on orderly transitions is not a good one. But there is at least one example from the UK of a good outcome that we can learn from. In 2023, its final operating coal plant—Ratcliffe-on-Soar—was shut down. While there were a few job losses (most through voluntary redundancy), the majority of the several hundred employees were redeployed elsewhere in the company, or retrained to work with partner companies in the energy industry.[5] This can only be done with a clear and early plan. If countries or industries wait until a few months before closure, it's too late. They are walking communities off a cliff. What's mad is that they can't claim to be surprised by the cliff edge; the closure of coal, oil, and gas industries is in the climate commitments we promise the world year after year.

Something to keep in mind is that while the *number* of jobs in energy is probably going to increase—and many workers will transfer between industries—some of these jobs are going to be less lucrative than roles in the fossil fuel industry. A wind turbine installer gets paid less than an offshore oil worker. That's not going to be an easy move for fossil fuel workers, and we should be honest and clear about what this change means for wages.

It's also important to keep in mind that energy industries were already in decline—before the rise of clean energy—because of productivity gains. Take coal mining in China as an example. Its coal production has skyrocketed since 2000. The number of miners "should" have doubled—rising from 5 to 10 million—to produce that amount of coal. But the number of miners has halved to 2.5 million. The average worker produces four times as much coal as they did in 2000 because a lot of their work has been replaced by mechanization and automation.

When people are automated out of a job, they rarely get support to find a new one. The transition to automation has been almost accidental and most people have accepted it and muddled through. The energy transition we're building is an intentional and controlled one. That's why it's viewed differently from job losses through productivity: People see it as a deliberate choice to destroy their livelihoods. But the fact that the world is planning for a transition means we *can* do it thoughtfully, providing support for workers who lose existing jobs.

III

RENEWABLE ENERGY

There is a lot of skepticism about whether we can run our lives on renewable energy, and even whether it's better for the environment in the first place. Renewables are far from perfect. How much energy they produce depends on how windy it is, or the time of day. We need somewhere to put the solar panels and wind turbines, which can create conflict with local communities and threaten biodiversity. They need minerals to build and generate waste, like most energy sources.

These are all challenges that we need to confront head-on, but they are not unsolvable. While renewables might not power 100% of our economies in the future, there's little doubt that they're going to be a huge part of the solution. They're one of our most powerful tools against climate change, so we'd better get building.

QUESTION 14

Don't solar and wind emit lots of carbon when we include the materials to build them?

Answer: Solar and wind are responsible for some carbon emissions during manufacturing, but *far*, far less than fossil fuels.

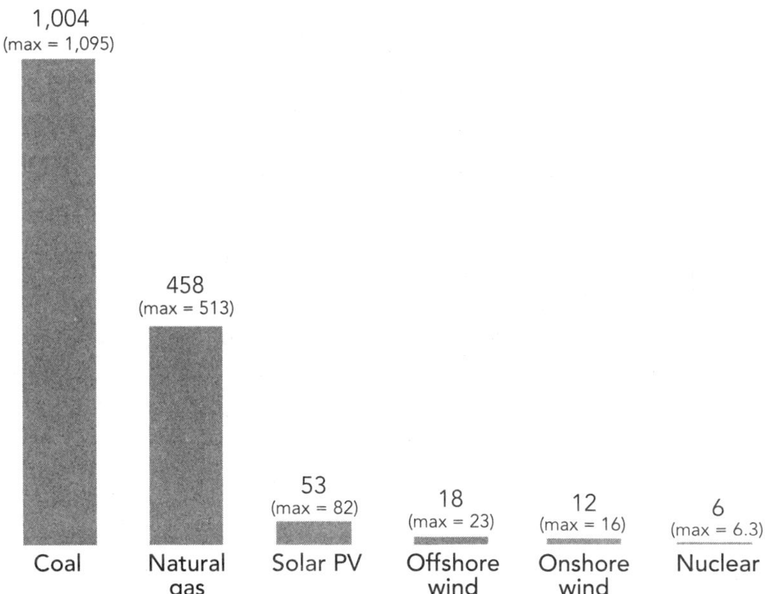

Figure 14.1

The carbon footprint of different energy sources: life-cycle greenhouse gas emissions, measured in grams of CO_2e per kilowatt-hour. This is shown as the median, with the maximum in brackets. Renewables and nuclear are very low-carbon even when we account for mining and manufacturing.

Data source: United Nations Economic Commission for Europe, 2021.

All of our energy sources have a carbon footprint. Coal and gas emit CO_2 directly when we burn them. Nuclear and renewable energy don't, but they still cause emissions during the manufacturing stage. Digging minerals out of the ground, processing them into usable components, and building the panels and turbines takes energy, and a lot of that still comes from fossil fuels. That's why we call them "low-carbon" energy sources, not zero-carbon ones.

But the argument that they emit *lots* of carbon—or are worse than fossil fuels—is nonsense.

Researchers can use a process called "life-cycle analysis" to calculate the total carbon footprint of different products, accounting for any greenhouse gases that are emitted along their supply chains, including any emissions from mining, transportation, and manufacturing.[1]

As you can see in figure 14.1, solar and wind emit some carbon, but *much* less than fossil fuels.[2] Per unit of electricity, solar emits 19 times less than coal. Onshore wind emits 84 times less. Nuclear, 180 times. It's not even a close race.

Where panels and turbines are produced matters: If we're building them in China where a lot of electricity is still coming from coal, the emissions will be higher than producing them in Europe, where the electricity is cleaner. But even the "maximum" emissions you'd get from solar and wind are *much* lower than fossil fuels.

Another way to look at this is to calculate how much carbon will be emitted to move our electricity systems from fossil fuels to cleaner sources. One study looked at 75 different mixes of technologies that the world could build up to 2050 to keep global warming below 1.5°C.[3] In almost every case, the total emissions, spread over decades, were less than 20 billion tons of greenhouse gases. That's around half of what we emit *every year* right now. While these emissions are not negligible, they're very small compared to the amount we'd emit with an electricity grid run on fossil fuels.

Another important point is that the carbon footprint of solar and wind will fall over time. The reason they have any emissions at all is because we're using fossil fuel energy to mine and manufacture them. As we decarbonize, our energy system will become cleaner. We'll start to manufacture solar panels using energy from solar and wind, not coal or gas.

WHAT WE NEED TO DO

The "carbon footprint" argument shouldn't stop us from building renewables. But there are things we can do to reduce their climate impact.

- **Produce solar panels and wind turbines in countries with lower-carbon electricity grids already.** A solar panel produced in France or the United States has a lower carbon footprint than one produced in China. Building supply chains in countries with cleaner electricity grids reduces their impact. We need to be careful with trade-offs here, though. Solar panels from China are cheaper, which means they're more likely to get built. The extra carbon emissions from them being produced in China might be worth it if it means we can deploy solar twice as much, twice as fast.
- **Electrify and decarbonize manufacturing within supply chains.** We'll need to be able to manufacture steel and cement, and process minerals using clean energy eventually. We're making some progress on this, but still have some way to go. Investing in innovative technologies now will make sure we have solutions scaled up ahead of 2050 so we're not building renewables with fossil fuels forever.

QUESTION 15

What happens when the sun doesn't shine and wind doesn't blow?

Answer: Solar and wind power don't produce electricity all the time, but we have options to store energy and balance it out to keep the lights on.

I'll let you in on a secret. The engineers who design wind turbines know that the wind doesn't blow all the time. Solar panel engineers know that daytime is followed by night. The National Grid—which keeps our networks running without the lights going out—knows both of these facts, but is still bullish on renewables.[1]

The fact that renewables don't produce consistent power all the time *is* a genuine problem. People don't just want cheap electricity: They want and need cheap, *stable* electricity. We can't accept a grid system that is vulnerable to blackouts. There will be periods when our supplies of sun and wind are out of sync with the country's electricity demand. Sometimes they'll be producing *more* than we need and at other times too little. What we need, then, is a way to *store* this energy.

To pick the right solutions, we have to think about how long we'll need to store energy for. We're not going to be short of options.[2] What's key is picking the right solution for a specific context and country. The challenge for the US is not the same as the challenge for the UK or Australia. Let's look at the solutions over three time frames.

Minutes to hours is the easy bit. Batteries are leading the way. They're very efficient at storing and releasing energy: You get back about 98% of the electricity you put in.[3] The downside is that they can only store and release small amounts at a given time. So, they're perfect for this hour-by-hour timescale but would be far too expensive to store energy for months.

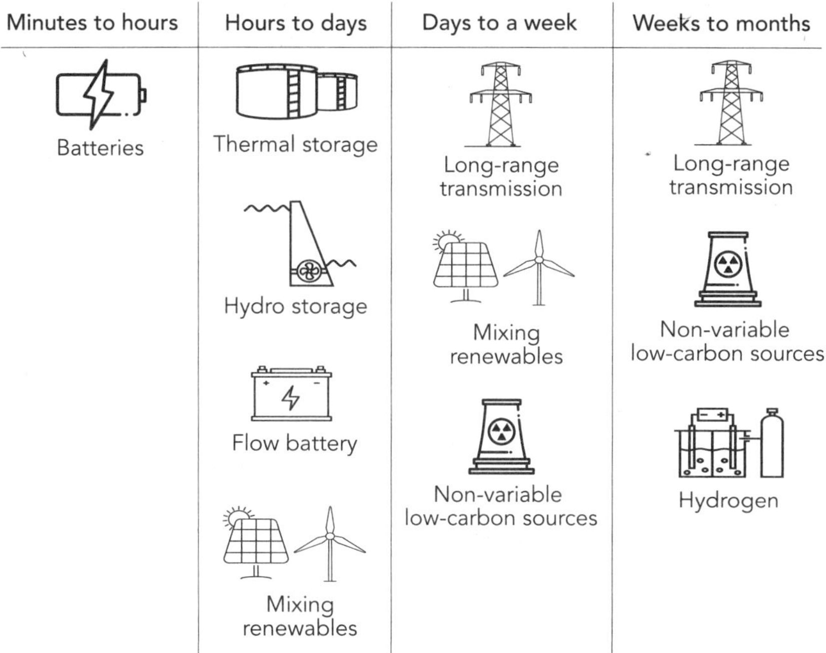

Figure 15.1
Options to store energy or balance our electricity grids: a nonexhaustive list of energy storage and grid balancing options over different time frames. Some of these methods can span multiple periods.

Balancing fluctuations over **hours to days** will be a bit harder. A simple solution of just solar and batteries might work for very sunny locations near the equator, but most countries will need a mix of solar and wind. What works out surprisingly nicely is that solar power is abundant in the middle of the day and in the summer; wind power is stronger during the night and in the winter. But adding other energy sources—such as hydropower, geothermal, and nuclear power—will help a lot too.

And we can take advantage of the fact that it's always sunny or windy *somewhere* in the world. You could, for example, build transmission networks across the US, balancing supply and demand from the West to the East Coast. You could do the same across Europe: Sunny Spain could help the UK when it's cloudy, and the UK could return the favor when it's windy. Some countries are even investing in transmission *across* continents.

Europe is looking to import electricity from large solar farms in Morocco, for example.

Again, when it comes to storing energy, there are several options. Other types of batteries, such as iron redox flow ones, can store and release power over slightly longer periods than the lithium-ion batteries you'd find in an electric car. Countries with mountainous or hilly regions can use pumped hydropower. When it's sunny or windy, water is pumped up to a high reservoir, where it's stored as potential energy; then when resources are scarce, this energy is released as water falls back down to a lower reservoir. Pumped hydropower is not quite as efficient as batteries—you only get around 80% of your electricity back—but it is cheap.[4] In most cases, this is the cheapest way to store energy over a week.[5]

The final option is heat storage. When electricity is abundant, we can store it as heat in various materials, which can be released later when supplies are low. There are now endless proposals for what could be used: water, rock, sand, molten salts. This is perfect for many industrial companies that run very high-temperature processes to produce steel, cement, and glass.

The biggest challenge will be long-term storage over **weeks to months**. Many of the solutions just explored will help here, but hydrogen storage is another option.[6] When electricity is abundant, it can be used to produce hydrogen, which can then be used as a fuel at a later date. The downside is that hydrogen is much less efficient than batteries. Only around 40–50% of the initial electricity can be used later. You are, effectively, throwing half of it away.

The energy analyst Chris Goodall calculated the cost of energy storage for the UK if it were to play it safe and keep two months' worth of electricity backup over winter.[7] Batteries would cost billions to tens of billions. Still a lot of money, but far less expensive than battery storage. Hydrogen might cost around $7 billion per year. A decent chunk of money, but far from crippling for the UK economy.

The world is still trying to work out the best path for long-term energy storage, and in that sense we're in uncharted territory. But we do have options. As countries start to run on more and more solar and wind, they will need to start committing and investing in their favorites.

WHAT WE NEED TO DO

For many countries, getting to 80%, 90%, or even 95% of clean electricity could be relatively easy. The final piece—giving us reliable supplies all the time—is the tricky bit. But there are things we can do to make it easier.

- **Find the right solution for each country.** There's no universal solution here. The mix of technologies that countries need will depend on how much sun and wind they get; how consistent this is across the year; who their neighbors are; and many other factors. Countries need to find the best solution for them.
- **Collaborate with neighbors.** Countries have traded electricity for a long time. France sells electricity to Germany, and vice versa. The US is in prime position to do this, and does so already: Maryland, for example, imports electricity from Virginia and Delaware. These neighbor relationships will be even more important with a low-carbon energy system. There will be times when one country has lots of solar and wind power, and others have very little. Trading across countries is an efficient way to balance out all these peaks and troughs. The alternative is that every individual country or state would have to overbuild energy sources to make sure it has enough in almost any scenario. That's a waste of resources. Maintaining these neighborly relationships will be a key geopolitical challenge to make sure that countries have secure energy supplies.
- **Get to 100% clean power soon to show that it can be done.** As I've said, long-term storage is a problem that isn't quite solved yet, which makes people—quite rightly—skeptical that it can be done. It's important that some countries, ideally the richest that can invest more in early innovations, get there quickly to break boundaries and prove that it can. Others can then learn from their wins (and mistakes) to get there faster and cheaper too.

QUESTION 16

Aren't renewables too expensive?

Answer: The costs of solar and wind have plummeted and are now cost-competitive or even cheaper than fossil fuels.

> Solar power is by far the most expensive way of reducing carbon emissions. . . . Wind is the next most expensive.
> *The Economist*, 2014[1]

> Solar power is now "the cheapest electricity in history."
> Report from the International Energy Agency, 2020[2]

These statements show how rapidly the costs of solar power have plunged. They've fallen 10-fold since 2010. The costs of wind power have plummeted too. Ten to fifteen years ago, fossil fuels *were* cheaper than renewables, but they've recently lost their price advantage. Figure 16.1 shows this clearly. Renewables have got much cheaper because they follow a "learning curve": as we build more of them, the prices fall. Fossil fuels have no learning curve (in fact, they never had one) and are not going to get much cheaper.[3]

In 2023, 96% of solar photovoltaic (PV) and onshore wind had lower generation costs than *new* coal and gas plants.[4] Three-quarters of these solar and wind plants produced cheaper power than fossil fuel plants that had *already been built.*

Of course, electricity grids running on solar and wind will need energy storage for when the sun isn't shining and the wind isn't blowing. The good news is that the costs of batteries are plummeting too; they've fallen around

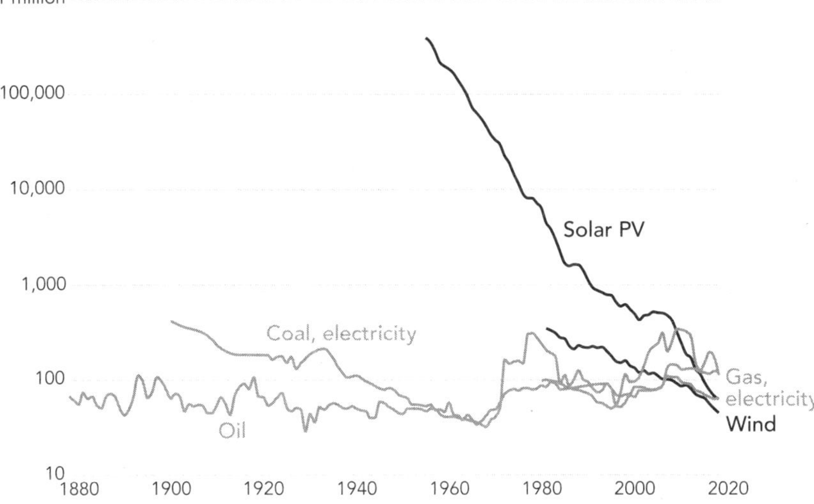

Figure 16.1

The cost of fossil fuels versus renewables over time: Renewable technologies beat fossil fuels on cost. The costs of solar and wind power have fallen dramatically. Fossil fuels do not show any learning curve of reduced costs, only volatility from year to year.

Data source: Way et al., 2022.

85% in the last decade. And there is no sign of this stopping; analysts think that prices could fall by another 50% by 2030.[5]

So when we look at the costs of solar plus *short*-term storage—up to five hours or so—solar and batteries can beat gas plants in many parts of the world.[6] This is not yet true everywhere, but in sunny countries, the combination of solar and batteries is already winning. In Australia, solar panels and batteries are estimated to be 30% cheaper than gas plants during peak demand.[7] In India, they are already cheaper than many existing fossil fuel plants.

But even these comparisons underplay the price advantage for renewables because fossil fuel costs are artificially low.

Two major costs have been missing from the price tag of fossil fuels: damage from climate change and from local air pollution. When we burn fossil fuels, they emit small particles and gases which we inhale and which damage our health. Air pollution is responsible for millions of premature deaths each year from cardiovascular disease, cancer, stroke, and respiratory illnesses.[8] In many polluted cities, people lose years of healthy life expectancy. Some estimate that these hidden environmental and health costs add up to almost $6 trillion a year globally.[9]

What's particularly damning is that these costs often don't fall on those who have caused the pollution. They will be passed on to future generations, and some of the world's poorest, who have contributed almost nothing to the problem. It's like running up a large debt with the expectation that someone else is going to pay it for you.

WHAT WE NEED TO DO

How do we take full advantage of the increasing cost benefits of renewables?

- **Deploy solar, wind and batteries faster.** These technologies follow learning curves, so the more we build, the cheaper they get. It's not rocket science.
- **Support poorer countries with the capital costs for renewables.** Solar and wind are being built in some emerging markets, but not as quickly as we might hope. Part of the issue is how these technologies are financed: Nearly all the costs come upfront, rather than being spread over decades. As I've already pointed out, if we want this to change, foreign investors need to provide finance at much lower and fairer rates. This is a massive barrier to the development of low-carbon markets in poorer countries and ultimately forces them down the route of fossil fuels, where the costs are spread over much longer periods. Richer countries can support poorer ones by offering concessional loans with low- or zero-interest rates; partnering with other countries or multilateral organizations like the World Bank to share the costs; and innovating on financial mechanisms that reduce the risks for investors.
- **Add a carbon tax.** Put a price on carbon so products that emit a lot of carbon dioxide get much more expensive, people buy less, and they switch to cheaper, low-carbon alternatives. The US does not have a federal carbon tax, although many states have implemented market solutions

such as a cap-and-trade scheme that try to produce similar outcomes. But many other countries are already doing this: The European Union has a carbon trading scheme, as does China; and countries from the UK to New Zealand, and Argentina to South Africa have a carbon tax on at least *some* carbon-emitting sectors. More than a quarter of the world's emissions are covered by a carbon tax or trading scheme.[10] So while a *global* carbon tax might seem elusive, lots of countries are already taking steps in the right direction and can be models for others to follow.

THINGS TO BEAR IN MIND

Energy sources are often compared using a metric called the "levelized cost of electricity" (LCOE). This estimates the cost of producing one unit of electricity over a power plant's lifetime. It takes the construction costs and running costs into account. In most countries, the LCOE of solar power and onshore wind is now lower than coal, gas, and nuclear.[11]

LCOE is not a perfect measure of energy costs. Costs can vary across the world depending on fuel prices, wind resources, and how sunny your location is. It also doesn't account for the total costs of the electricity system, which depends on the mix of sources you're building, not just one on its own. Calculating total system costs is more complex and will vary depending on the mix of electricity sources that countries have.

One argument people make is that solar and wind power can only compete because they get lots of subsidies. This was true in the past, but it's not today. Some countries gave generous subsidies for renewables when they were just getting off the ground and were much more expensive than they are today. Renewable energy producers in early movers like Germany and the Netherlands were given long-term subsidy contracts over several decades and are still benefiting from them today. These will fizzle out once those plants age and come to the end of their lives. Many new plants today are economically feasible without any subsidies at all. The reports on LCOE above are all based on unsubsidized costs. A 2023 study expects that in the coming decades, solar power will dominate electricity markets simply because it's so cheap.[12]

It's also worth pointing out that fossil fuels receive subsidies too. Governments support the production of fossil fuels and help consumers by subsidizing fuel, and electricity, or reducing taxes on energy. These subsidies totaled $1.3 trillion globally in 2022. This was much higher than the years before—in 2020, it was $500 billion—because of the surging costs of fossil fuels after Russia invaded Ukraine.

These subsidies don't include the costs of damages that come from burning fossil fuels. There is an obvious solution that takes account of these hidden costs: a carbon tax (which I recommended previously). It "puts a price on carbon" so that the damages from climate change *are* reflected in its market price. The problem is that there's no agreement on what this price should be. We don't know exactly how much climate change will cost us and we don't all agree on how to "value" a lost life.

An international analysis of almost 150 studies found average costs between $132 and $283 per tonne of carbon.[13] In other words, a product that emitted a tonne of carbon dioxide to generate should have $132 to $283 added to its price. But there is a huge variation in these estimates, even in a single country. Barack Obama's administration estimated this price to be $43 per tonne.[14] Donald Trump's administration slashed it to just $4 per tonne. Joe Biden's put it at $51, but his Environment Protection Agency talked about upping it to $190.[15] That's a 50 times difference between subsequent administrations in the US alone. Try to get an agreement among every country in the world on what this price should be. So, while the "missing" costs of fossil fuels point is crucial, you could end up in endless cycles of debate about *what* those costs should be. Nevertheless, the fundamentals remain unchanged: Renewables are getting cheaper, while fossil fuels are not.

Finally, while the costs of adding a new solar or wind plant are pretty cheap, what *could* get quite expensive on a renewables-only electricity grid is filling the final 20% of generation. Deploying more solar and wind when they're at 10%, 20%, 30%, or 40% of the electricity mix should continue to make electricity cheaper—at least if it's priced based on the cost to produce it. But these reductions could eventually run out at much higher penetrations. One study, for example, found that solar and wind pushed down costs significantly until they made up around 60% of the electricity mix.[16] That's when costs reached their low point. Then costs slowly crept up again, but still stayed well below the cost of fossil fuels. Getting the final 10% of stability in the grid was expensive from these technologies on their own. That's why many countries will need a portfolio of sources. Solar and wind could make up the bulk of the electricity grid but the energy transition will be easier and cheaper if it's supplemented with some nuclear, hydropower, geothermal, or other low-carbon sources.

QUESTION 17

Don't solar panels and wind turbines generate huge amounts of waste?

Answer: Solar and wind generate far less waste than alternative energy sources like coal and much less than other waste like plastics or electronics.

One of the most popular takedowns of solar and wind power has no explanatory text or numbers. The worst kind. Someone just shares a picture of solar panels being landfilled or turbine blades being dumped in a field. Everyone shakes their head, thinking they've been duped into believing that these technologies are "green."

No doubt about it: This waste is a problem, and the world needs to do better at repurposing and recycling solar panels and turbines. But the images seem worse than they are because they're missing the context of how much waste is produced by other energy sources.

The chart shows how much waste would be generated per megawatt-hour of electricity generation for solar, wind, and coal. Natural gas isn't included here since it produces very little *solid* waste, which gives it an advantage over other fuels. Though it does produce some rock waste during the mining process, which is discussed in Question 34.

Coal generates almost 50 times more waste than solar, and 500 times more than wind. Most of the waste from coal comes in the form of coal ash. For solar, it's the panels. The blades for wind. But people don't share pictures of massive heaps of coal ash, just pictures of piles of used turbine blades or panels. Moving from coal to renewable energy will reduce waste, not increase it.

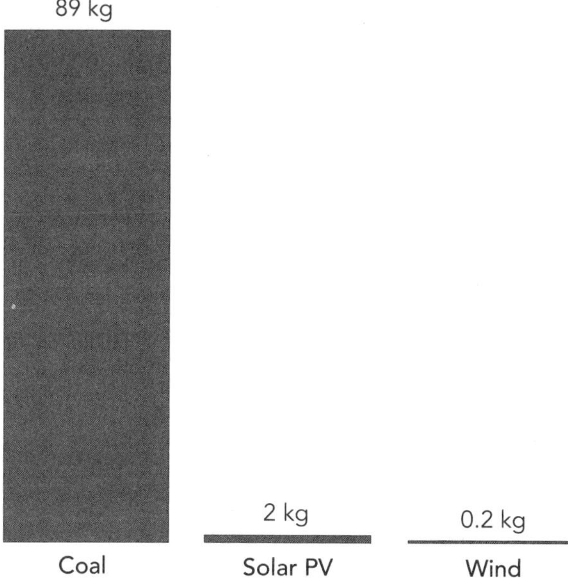

Figure 17.1

Waste generated by different electricity sources: waste generated per unit of electricity production, measured in kilograms of waste per megawatt-hour. Waste for solar and wind is in the form of panels and turbine blades. Coal is mostly ash.

Data sources: Calculated based on industry figures; Center for Sustainable Systems, University of Michigan; David Osmond; CleanTechnica; and the EIA.

A recent research paper estimated that globally we produce and manage approximately the same mass of coal ash *per month* as the amount of solar panel waste we expect to produce over the next 35 years.[1]

We can also compare solar and wind waste to other types of waste, like electronics, plastics, or household waste. To do that, let's look at how much waste the average person would generate over the next 25 years (the average lifespan of a wind turbine; solar panels can have a slightly longer lifespan of around 30 to 35 years).

I'm going to assume that I'm like the average person in the US. I have calculated how much waste would be generated from different electricity sources if I got *all* my electricity from that source. So, imagine that I got all my electricity from solar PV over the next 25 years. Or wind, or coal. That's unrealistic, but it's a good stress test of how the different technologies stack up.

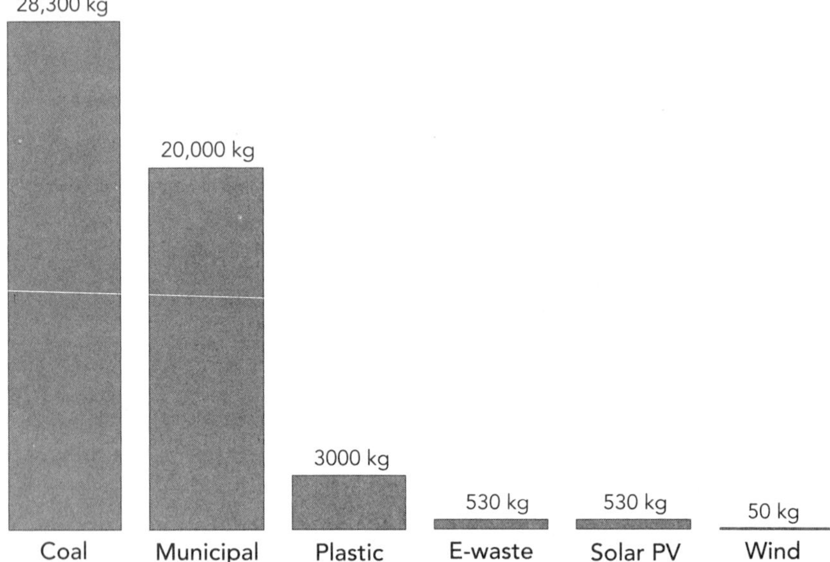

Figure 17.2

Renewable energy versus other types of waste: waste generation over 25 years for the average person in the US. Electricity waste assumes that all of a person's electricity came from that one source (e.g., all from coal or all from solar PV).

Data sources: Calculated based on industry figures; Center for Sustainable Systems, University of Michigan; David Osmond; CleanTechnica; EIA; World Bank; OECD; Global E-waste Monitor.

I'd generate 28 *tonnes* of coal ash, 530 kilograms of solar PV waste, or 50 kilograms of wind blades. This is shown in the chart. Beyond energy, in other parts of my life, I'd generate 20 tonnes of municipal waste, 3 tonnes of plastic, and 530 kilograms of electronic waste.

Electricity consumption in the US will probably increase in the next few decades as we electrify transport, heating, and other industries. That's fine: Even if we double the solar and wind totals, renewable waste will still be small compared to other types of waste.

WHAT WE NEED TO DO

Even though solar and wind generate less waste than people are led to believe, there are things we can do to help further.

- **First of all, stop basing your worldview on random photographs without context.** You can't understand the world's energy and waste systems this way.
- **Make the final 20% of wind turbines–the blades–recyclable.** This will not just reduce the amount going to landfill but make resource use more circular too. We'll need to manufacture fewer new materials. There is a lot of research and development in this area.[2] The manufacturer Siemens has launched its "RecyclableBlade," which is ready for commercial use. Many people have called for Europe-wide bans on blade landfilling this decade. I think recyclable blades will become a reality.
- **Make sure solar panels are recycled.** Most of the bulk materials in solar panels—such as aluminum and glass—can be recycled. Unfortunately, recycling rates in many countries are pretty low (although reliable numbers on this are often hard to find). Investing in technologies that make this process easier and bringing in regulations or policies that require higher rates of recycling could speed things along. Most of the solar panels that have been installed won't come to the end of their life until a decade or two from now. The aim is to make sure we have widespread panel recycling across the world by then.

THINGS TO BEAR IN MIND

It's not just the amount of waste that matters, but its toxicity too. A kilogram of arsenic or mercury is not the same as a kilogram of cotton in municipal waste. How concerned should we be about the toxicity of waste from solar or wind?

Let's start with wind. Around 80% to 85% of wind turbine components can be recycled; it's the blades that can't be (at least not easily).[3] In the US and Europe, blades are categorized as nonhazardous waste and can be sent to landfill. The risks to human health are extremely low.

What about solar? Despite critics thinking they contain various toxins, the only health concern from solar panels is the small amounts of lead in silicon panels, and trace amounts of cadmium in cadmium telluride (CdTe) ones.

Cadmium is in CdTe panels in very low concentrations—only around 0.1% by weight—and they are currently collected, with the cadmium used in new modules. The key thing is to make sure that these panels are collected appropriately at the end of their life, so that the cadmium can be properly managed.

A final point on the toxicity of energy waste. Coal ash—which we've been producing en masse for centuries—contains elements such as mercury, arsenic, lead, cadmium, and chromium. So, moving from our current energy system to renewables will not only reduce the *amount* of waste that's generated; it will also reduce the toxicity. This is why we need to have these discussions based on data, not viral images on social media.

QUESTION 18

Won't we run out of land to use for solar panels and wind turbines?

Answer: We might need to use a bit more land for energy than we do today, but it will still be just a small percentage of total land.

Many people worry that our planet will turn into a giant wind or solar farm. Newspapers publish images of solar panels stretching for miles across the fields. But pictures can be misleading. They don't tell us how much power those panels are producing or how much land the alternatives need. Let's put the numbers on it.

We can compare energy sources based on how much land they need *per unit* of electricity production. Here we need to count not just the land used by the plant—the solar panel, the wind farm, or the coal power station—but all the inputs that go into it. Land is needed to mine minerals and fuel, to build roads to service power stations, provide a connection to our electricity grids, and to handle the materials from the power plant at the end of its life.[1]

The results are in figure 18.1. Nuclear is the most land-efficient way of making power. Next is gas, which is also pretty land-efficient. Putting solar panels on roofs is a great way to save land; its footprint is just the land needed to mine the materials. Utility solar—solar plants on farmland, in deserts, or on other terrain—needs about 50 times as much as nuclear. This seems like a lot until you consider that coal plants can be almost as land hungry. In some studies, coal needs *more* land than solar because the mining footprint of coal is so high. Wind can use very little or a lot of land depending on how you look at it. The *direct* land use of the wind turbines themselves is tiny. The *visual* footprint is not. More on this distinction in a minute.

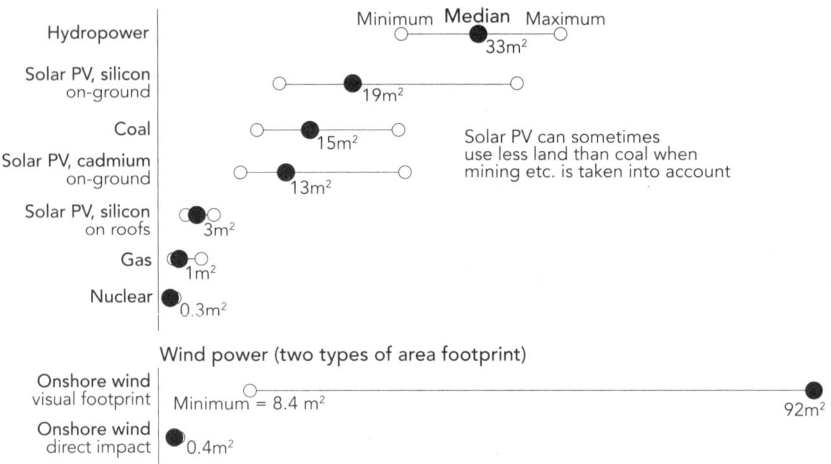

Figure 18.1

How much land do different electricity sources use per unit of power? Land use is measured in square meters per megawatt-hour (MWh) of power. This is calculated over the full supply chain, so includes the land use of mining, transport, and other impacts.

Data source: United Nations Economic Commission for Europe, 2021, and author calculations for wind.

Comparing land footprints per unit of electricity is fine, but we need some context on how much land the world would need to be powered by low-carbon energy. Let's start by imagining the world is powered by nuclear. Nuclear could power the world on less than 0.01% of land.

Solar would need just 0.5% of land.[2] Countries at higher latitudes—such as the UK—would need more; those in the sunny tropics a bit less. But total land use in most countries could be small, especially if we make good use of rooftops.

Wind is different. The *direct* land use of the turbines, roads, and other infrastructure is tiny. It's just the small amount of land where the turbine is stuck in the ground. You could power the world using less than 0.1% of land. But what people mean when they talk about the land use of wind power is its visual footprint: It's the turbines they can see. If we count all this space between turbines as "land use," we'd need 2% to 3% of the world. In some countries—especially small, densely populated ones—wind turbines would stretch over 10% to 15% of land, if not more.

With wind power, the land is not really being *used* when it comes to the visual footprint. Most turbines will be placed on farmland. The farmer or landowner gets some additional income from the power, but they can continue to grow crops between the turbines as usual. The "land use" is still farming even if the visual footprint is wind blades. This is not the same for other technologies such as a coal or nuclear plant: The land can't be used for anything else.

Finally, it's worth thinking about what we already use our land for.

Half of the world's habitable land is used for farming. Energy production would never come close to this figure. Sure, we need to eat, but a lot of this land is not used for food production. The US uses 2.5% of its land—an area the size of the UK—to produce bioethanol, which provides 10% of its motor gas supply. If it used that land for solar farms instead, the US could generate enough electricity to power itself two to three times over. Yes, you read that right. That's probably the amount of electricity that the US will need in the future once it has electrified all its cars, heating, and industrial plants. So it *does* have enough available land; it just needs to decide what to do with it.

Why do we hardly bat an eye at this land being used to produce a small amount of biofuels, but shudder that a solar farm of that size could power the entire US and then some?

WHAT WE NEED TO DO

As individuals, we can't change the physics of how much land is needed to site a wind or solar farm. But there are a few things we can do to make energy in a land-efficient way *and* make sure that the concerns about land use don't hold us back in the energy transition.

- Put solar panels on your roof, if you can afford to. This generates power using barely any land at all (only the land that's needed to mine the materials). Rooftop solar often sets off a ripple effect of adoption: Once a few households in a village or street get one, others copy.
- Ignore the provocative headlines and clickbait images of fields full of solar panels. If they don't have context on the numbers, they're there to provoke you, not inform you. See it for what it is: lazy journalism.

- Avoid repeating the same old tropes about how renewable energy has ruined our landscapes. It has become something of an obligation to comment on how ugly turbines are when you pass by a wind farm. I'm not actually convinced most people believe that. But how we talk about these technologies matters—it shapes the culture around them—so if you *aren't* offended by the sight of some turbines and solar panels, say so. Make it okay to support them.

At a societal level, here's what we can do.

- Get solar panels on as many redundant surfaces as possible: homes, commercial buildings, parking lot roofs. They aren't being used for anything else, so why not generate power close to where people need it? There are two things to note here, though: Rooftop solar is often more expensive than utility-scale farms, and solar on roofs won't give us enough power for everything. We'll still need larger solar farms in fields and deserts.
- Invest in offshore wind. This avoids some of the constraints of onshore wind (although it'll still be in the eyeline of some coastal and island communities).
- Develop partnerships where local communities benefit from having renewable energy farms close by. Some people *are* rightly concerned about changes to their local area, especially in areas where wind resources are abundant. This is a problem if the local communities don't see any of the financial benefits of these programs, not even in terms of lower energy costs. Developing co-owned projects with communities or offering subsidized electricity rates could make this transition fairer.

THINGS TO BEAR IN MIND

There are a few exceptions–mostly very small countries—where renewables won't fit. If South Korea was to run on wind power, it would need to use two-thirds of its land. Even if Singapore covered every inch of land with solar panels, it wouldn't be able to meet its electricity demand. These are the exceptions, not the rule, but it's important to highlight that solutions need to be tailored to the local context.

It's also worth looking at the data on how much land an energy system with a *mix of different energy sources* will need in the future. To get *all* our power from solar, wind, or nuclear is not going to happen (nor would it be a good outcome). A bunch of studies have estimated land-use needs in realistic low-carbon energy systems. One mapped out five possible and affordable scenarios to decarbonize the US economy by 2050.[3] It estimates that the US will need 1% to 6% of land for solar, and 5% to 22% for the visual footprint of wind.

Some parts of the European Union have fewer solar resources because they're at a higher latitude. But even there, it's possible to meet demand from renewable sources. It could meet its net-zero goals by 2050 using 3% of land for solar farms and up to 15% of land for wind energy.[4] That's the total visual footprint. Converting just 1% of land to solar farms could already meet its electricity needs *today*.

QUESTION 19

Can we build electricity grids fast enough?

Answer: Connection queues are one of the biggest barriers to rolling out renewables, but there are things we can do to speed things up.

When we turn on the lights, we tap into a finely tuned electricity grid. These grids connect where electricity is *produced* to where it is *consumed*. They carry electricity across vast networks via cables, often over long distances. Grid operators need to make sure that power plants produce exactly the amount of electricity that consumers want, minute by minute and second by second. When you think about it, it's incredible that our grids work so seamlessly.

One of the challenges of this energy transition is that we want to add lots of new electricity plants at once. The risk is that this makes our electricity grids unstable and overloads them, which is why they're being added painfully slowly.

Queues on the electricity grid are building up across the world. Without a grid connection, new plants are stuck on hold. Globally, "advanced" solar and wind projects in the queue—projects that are ready to go and waiting patiently—represent *five* times the amount of solar and wind that came online in 2022.[1] Can you imagine how much further ahead we'd be if the world deployed solar and wind five times as fast?

In the chart, you can see the rapid rise in the amount of renewable energy projects in the United States that are waiting for a grid connection. The queued projects represent twice the power of the entire US electricity grid.[2] Renewable energy developers will need to wait at least five years for a connection.

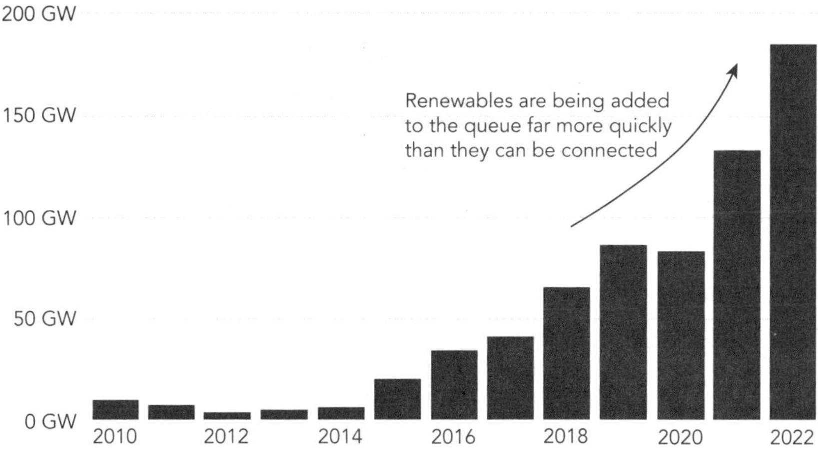

Figure 19.1

The growing queue for grid connections in the US: The amount that the grid connection queue in the United States is growing every year, measured in gigawatts (GW).
Data source: International Energy Agency.

The UK isn't much better. Ofgem, the country's electricity regulator, estimates that the amount of projects sitting in the connection queue will deliver four times the power we will need by 2050 to decarbonize our grid.[3] But we can't build transmission and distribution networks fast enough. There are around 600 new project applications every year in England and Wales.[4] Developers applying for a grid connection today are being offered a connection date in the late 2030s.

If anything, rich countries are slower at building out their grids than poorer ones. India and China, for example, take around two years to build a transmission line (which are the cables that carry electricity from a power plant to your home).[5] In Europe and the US, it's four or five years.

What's going on? Why does it take so long to get a grid connection? Well, you can't just plug a solar or wind farm into the grid and go. Every time you add a new generator, there's some risk that it will pull the whole system down, causing havoc to thousands or millions of homes. So, it's the regulator's job to assess whether projects can be connected, and how they will fit into our intricate, carefully balanced system.

The process of approving new energy projects is set up for our old way of doing things, but rather than having a few big fossil fuel plants that run most of the time, we're building a system with hundreds (or more) solar and wind farms that are much more variable. This just doesn't work with the existing energy system. Another key problem is that transmission lines cover long distances, which require a lengthy planning process. To build a solar farm, you often only need permission from a small local community in one place. But grid networks can cut through hundreds of communities. Building a 340-kilometer grid line in Germany needed around 13,500 permits.[6]

WHAT WE NEED TO DO

There are no sexy solutions to these problems. They're changes that most of us will never notice. But here's what we can do to improve things.

- **Invest in grids properly; stop treating them like an afterthought.** Grids aren't just a bunch of pylons and cables we throw up after we've built solar and wind farms. They're the central machinery that keeps our electricity systems running. While investments in renewables have been increasing quickly over the last decade, money going into grids has stagnated. The world invests around $300 billion in grids every year; this needs to double to $600 billion by 2030.
- **Introduce simpler permitting rules.** Having to apply for thousands—possibly tens of thousands—of permits for a single line is not going to get us there. We need a simpler system with less red tape. Some countries are trying to make this happen: Joe Biden's administration in the US signed off on new rules that would speed the process up. While Trump has promised to accelerate permitting reform for energy infrastructure, it's mostly targeted at fossil fuel and nuclear plants. Solar and wind might, instead, have *higher* levels of scrutiny and bottlenecks. Some states might still be able to get renewables connected faster, but the lack of national-level support will probably still slow things down.
- **Train and upskill grid engineers.** Many countries don't have enough skilled workers to expand the network quickly enough. You will notice

that this is a recurring theme in the book: There is a huge gap in the labor market for people who can implement the solutions we need. As I've said before: fewer people like me talking about building stuff, and more people doing it.

- **Kick "zombie" projects out of the queue.** The UK is trying to overhaul the queuing system by eliminating "zombie" projects that are unlikely ever to be built. They are effectively scrapping a "first come, first served" strategy to make sure the most promising and advanced projects get priority. The government claims this will cut waiting times from five years to six months.[7] In the US, a similar transition is underway. For years, the country followed a "first come, first served" system, which led to long delays and crowded interconnection queues. In 2023, a federal order mandated a shift to a "first ready, first served" approach, requiring projects to meet key development milestones to advance. But despite the policy change, a full nationwide transition is—ironically—taking some time.
- **Innovate and invest in new grid technologies.** Different technologies might also relieve some of the pressure and reduce gridlocks.[8] Most transmission lines are made from a combination of steel, copper, and aluminum. The problem is that these tend to sag from heat, so operators must limit how much power can be passed through them. If these were replaced with carbon fiber, they would be much less vulnerable to overheating and could carry a lot more electricity. This means countries would need to build fewer new lines. It's an innovation that is happening, with start-ups working hard to bring it to market.

QUESTION 20

Don't wind farms kill lots of birds and wildlife?

Answer: Wind farms do kill some birds and bats, but just a fraction of the numbers killed by other threats we rarely talk about.

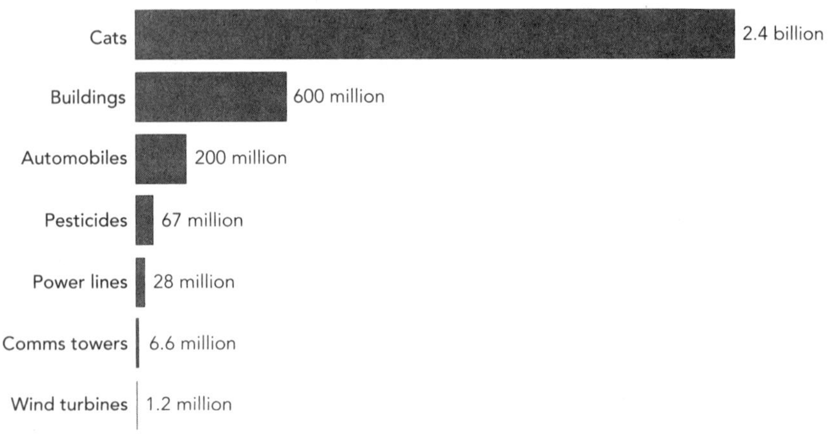

Figure 20.1
Number of wild birds killed by different hazards in the US each year.
Data sources: Loss et al., 2015, 2013; US Fish and Wildlife Service, 2012; Subramanian et al., 2012; American Bird Conservancy, 2021.

"Windmills . . . kill all the birds," claimed Donald Trump.[1] Wind turbines do kill birds, just not all of them. Not very many compared to other hazards like cats (which is sad for me to admit), buildings (they fly into them . . .), cars, and pesticides.

The chart in figure 20.1 gives a rough estimate of the number of birds killed each year by different threats in the United States. Wind turbines kill just over a million; maybe a few million at most. Cars, buildings, and pesticides kill tens to hundreds of millions each. Cats kill over two billion. Yet you don't hear Donald Trump talking about tearing down buildings, taking cars off the road, or making the public give up their cats. And this doesn't even include the 10 billion birds that Americans kill for meat every year.

But what *types* of birds are at risk matters just as much as the total number. Killing rare species is worse for biodiversity than abundant ones. Studies have shown that birds of prey such as eagles, raptors and hawks, and shorebirds and stork-like species are at much higher risk of collisions than other types, such as songbirds.[2] This is true for a couple of reasons. First, birds of prey and shorebirds will often use ridgetops to get lift from the wind. Naturally, this is also a good spot for wind turbines so they're sometimes fighting for that space. Second, some (but not all) of these birds are migratory; if wind farms are on their route, this puts them at higher risk. More indirect impacts of wind farms—which might not be reflected in death statistics—include their effect on the disruption of migratory patterns.

Studies have shown that bats are often at a higher risk than birds.[3] Estimates suggest each turbine kills 6 to 20 bats per year. So, while the total number of birds killed by turbines is low compared to other hazards, the threat to particular species and bats is more concerning.

WHAT WE NEED TO DO

We shouldn't just wave away complaints about the impacts of wind farms on wildlife because other hazards are more dangerous. We want to build them responsibly and try to minimize their impact as much as we can. There are plenty of things we can do.

- **Turn off wind turbines when wind speeds are low.** Bats tend to get hit by wind turbines when wind speeds are very low. They struggle to fly in windier conditions. We can prevent a lot of bat deaths by switching off our turbines when there isn't much wind. A study from a Pennsylvania wind farm found that bat deaths fell by 44% when wind turbines were turned on at a wind speed of 5, rather than 3.5 meters per second.[4] And they fell by 93% when this was increased to 6.5 meters per second. You might think that this would hinder energy supply and eat into owner profits. But studies suggest it doesn't make much difference. An Australian wind farm found it could reduce deaths by 54% while only losing 0.1% of revenue.[5] Another study in Cadiz found that bird deaths were halved with only a 0.07% loss in energy production.[6]

- **Don't put wind farms in high-risk areas for birds and bats.** As areas like ridgetops are a prime spot for migratory birds and raptors, we should try to avoid these if we can. Admittedly, that's not always easy—and sometimes involves a trade-off—as these can be a very good spots for wind farms.
- **Build fewer, larger turbines rather than many small ones.** This reduces fatality rates, as studies have shown that birds are more protected on a wind farm with a few large turbines than lots of small ones.
- **Paint the turbines black.** When birds get close to turbines, the blades spin so quickly that it blurs their vision. But they're much more visible if you paint the blades black. Some tests of this approach in Norway reduced bird deaths by more than 70%.[7] Crucially, the greatest reduction was found for raptors. No white-tailed eagle carcasses were found after the turbines had been painted.
- **Play alert noises to deter bats and birds.** High-pitched sounds (that humans can't hear) can deter some bat species. An experiment in Texas played these sounds and reduced the deaths of two species of bats by 54% and 78%.[8] Other systems can identify eagles in the nearby area, and either emit distracting noises or switch the turbines off automatically.
- **Use GPS to track and find the optimal height for turbines.** Surveillance can help scientists understand the flying patterns of migratory species. That means we can pick the optimal heights for turbines when they're being constructed. GPS can also alert wind farm generators that migratory flocks are in the area, so generators can switch turbines off during high-risk times.

THINGS TO BEAR IN MIND

Coming up with estimates for how many birds die from cars, cats, and flying into buildings every year is not an exact science. No one has gone around measuring the number of birds killed by cats across the entire United States. Instead, researchers study how many birds the average outdoor cat kills and multiply this by the total number of cats. Or they'll study the number of bird deaths from building collisions in a smaller area, then multiply by the number of buildings in the country. That means the numbers are quite uncertain: Some of them could be twice or half as high as I've shown in the chart. But they do give us an overall sense of scale and magnitude, and the ordering is correct.

You might wonder where the estimates for the number of birds killed by wind turbines come from. Researchers estimate that around 4 to 18 birds are killed per turbine per year.[9] If we multiply this by the number of turbines in the US, we get between 200,000 and 1.2 million bird deaths per year. In the chart at the top of this question, I took the highest figure.

Of course, we'll build more wind turbines in the future so this number will probably increase (unless we implement the strategies described above to reduce turbine collisions). Even if we multiply the world's wind power by five or 10 times what it is today, it's still a relatively small threat compared to other hazards.

Finally, it's important to consider bird deaths due to fossil fuels: Climate change, air pollution, acid rain, and toxic waste from coal mines also affect bird populations. These combined risks are likely to kill and disrupt bird species far more than wind power.

Keeping fossil fuels instead of switching to renewables to "save the birds" doesn't make sense. A striking testament to that point is that in the UK the Royal Society for the Protection of Birds built a wind turbine to power its own headquarters.

IV

NUCLEAR POWER

If the world's approach to nuclear 50 years ago had been different, we might be on a much better climate trajectory than we are today.

Many countries were building nuclear plants rapidly in the 1970s and '80s. The countries that went all in on nuclear—such as France and Sweden—are still home to some of the world's cleanest electricity grids. For decades, they've had flourishing economies while emitting much less carbon than their neighbors.

If the world had fully embraced nuclear power and stuck with it, we'd be burning far fewer fossil fuels and have much less polluted air than we do now. Unfortunately, nuclear power fell out of favor. Concerns about safety and radioactive waste, and challenges around costs and construction times, mean that nuclear power is shrinking in the global electricity mix.

There is still time for a comeback, but it will need a dramatic shift in public and political support, and major reforms to make it possible to build nuclear plants cheaper and faster.

QUESTION 21

Isn't nuclear power dangerous?

Answer: Nuclear power is not risk free, but it's one of the safest energy sources we have.

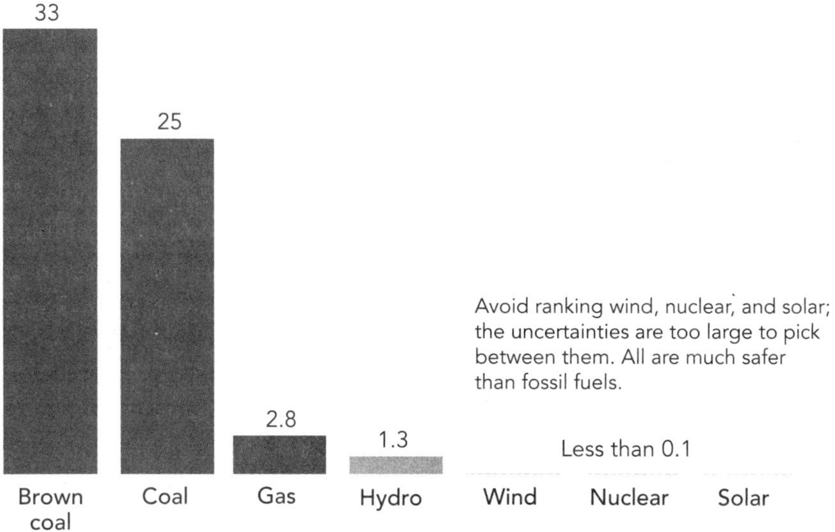

Figure 21.1

Deaths per unit (terawatt-hour) of electricity generated. This includes deaths from accidents, air pollution, and radiation. Brown coal is the lowest-grade coal, and particularly polluting.
Data sources: Markandya & Wilkinson, 2007; Sovacool et al., 2016; UNSCEAR, 2008, 2018.

Nuclear power is one of the safest energy sources we have, which is something I would not have believed 15 years ago. But look at the data and that's exactly what we find.

No energy source is completely safe. Each has some risk, whether it's sourcing or mining the minerals or the fuel, building the plants, running them, or accidents during their operation. Let's compare our electricity sources based on how many deaths they've caused. To do this, we can't just add up the total number of deaths for each source because we've produced a lot more power from fossil fuels than solar, nuclear, and wind power. Instead, we're going to compare them on how many deaths they cause *per unit* of electricity production. This includes deaths from pollution *and* accidents in supply chains.

The results are shown in figure 21.1 at the top.[1] What immediately stands out is how bad fossil fuels are. Especially coal: It's the dirtiest fuel by a long shot. Coal comes in various forms; some deposits have more impurities and create more pollutants than others when they're burned. "Brown coal" is a type that generates a lot of pollution, so it ranks worst. These numbers might even be too kind to fossil fuels: They're based on power plants in Europe, where air pollution controls are much tighter than in many other parts of the world.[2] Globally, the numbers are probably bigger.

When people say "Nuclear energy is unsafe," they are often oblivious to how it compares to fossil fuels. Nuclear energy results in more than 99% fewer deaths than coal, and around 98% fewer than gas. Keeping fossil fuels over nuclear because nuclear is "too unsafe" makes no sense.

WHAT WE NEED TO DO

How do we build public trust in nuclear power, and help people see the relative risks of different energy sources more clearly?

Close the perception gap between the dangers of nuclear power versus fossil fuels. This perception gap is caused by the fact that nuclear accidents are one-off, dramatic events. There have been only a few major disasters in nuclear power's history, but they are headline-grabbing, traumatic stories that are hard to look past. Nuclear disasters are made into blockbuster movies; deaths from fossil fuels aren't in the same way. They don't happen in single, one-off events. Instead, they kill thousands of people *every* day. Researchers estimate that fossil fuels cause 4 to 8 million premature deaths per year from air pollution alone. That's around 11,000 to 22,000 every day. You never see the headline "Fossil fuels killed 11,000 people from air pollution yesterday," yet a single accident at a nuclear plant that killed no one would be one of the year's most talked about events. We should put more focus on the ordinary, not the extraordinary.

Be honest about the risks of nuclear power. No energy technology is completely safe. There could be another disaster (see details about previous disasters in "Things to bear in mind"), especially if the world was to build a lot more nuclear plants. That's a real possibility. We need to focus on building the safest reactors we can and make smart decisions on where to site them. But "safety" is no excuse to stick with fossil fuels, because they are the least safe option we have.

Stop obsessing over the safety record of nuclear versus renewables. Whenever I share the chart included in this question, people inevitably focus on the wrong part. They zoom into the bottom-right corner and squint to see any differences between nuclear, solar, and wind. The reality is that there's little to pick between them. The quality of this data is not sufficient to rank nuclear, solar, and wind with any certainty. I would never say that, based on this data, solar is *the* safest, or that nuclear power is safer than wind. So please don't spread this argument either (people already do and quote me as the source . . .). All three are exceptionally safe compared to fossil fuels. That's what we need to focus on.

THINGS TO BEAR IN MIND

At this point, you might still be skeptical. How can the numbers for nuclear be so low? They don't match up with the image of nuclear energy that many have in their heads. So, let's take a closer look.

Our worries about nuclear energy are driven by memories of two key events: Chernobyl in Ukraine in 1986, and the 2011 Fukushima disaster in Japan. These are the only two events that reached level 7 (the maximum classification) on the International Nuclear Event Scale. Three Mile Island reached level 5.

How many people died in these events? Take a quick guess now.

In Three Mile Island, **zero**.

No one died during the Fukushima disaster either.[3] Forty to fifty people were injured from the blast or suffered radiation burns. In 2018, the Japanese government reported that one worker had since died from lung cancer because of radiation exposure from the event. This link is disputed by some, but I count it in the numbers presented above. Over the last decade, many studies have concluded that there was no increased risk of cancer for local populations. A difficult question is how many people died because of the *response* to the meltdown. More than 100,000 people in the surrounding areas were evacuated, causing physical and mental stress, and disruption of healthcare facilities. The Japanese government estimate that evacuation led to 2,313 premature deaths.[4] Again, people dispute that these should be included—it was, after all, also the aftermath of a major tsunami—but let's be conservative. That brings Fukushima's total to **2,314 deaths**.

Chernobyl was, by far, the most poorly handled nuclear disaster and it's the hardest to provide good estimates for. I estimate that we could say with confidence that it killed at least **300 to 500**. This doesn't include any evacuation-related deaths like those added for Fukushima, because it's impossible to find any numbers on this.

The *confirmed* death toll for Chernobyl is **less than 100**. Two people died as a direct result of the blast. Twenty-eight workers and firemen died in the weeks that followed from acute radiation syndrome. By 2006, five of the surviving plant workers had died from cancer-related causes, which could have some link to the disaster. Fifteen people died from thyroid cancer after being exposed as children to milk contaminated with iodine, in the days after the disaster.

From an in-depth look at the literature, I estimate that in the decades that followed there could have been several hundred more deaths because of thyroid cancer from radiation exposure. My mid-range estimate is 433 deaths.

There have been lots of claims about the impacts of radiation exposure on the rest of Europe. In the early 2000s, the World Health Organization (WHO) estimated that there might be an additional 4,000 deaths due to Chernobyl over the years. Many people now think this is too high, especially with an improved understanding of radiation exposure and little evidence of higher death rates in wider populations. In multiple follow-up reports, the United Nations Scientific Committee on the Effects of Atomic Radiation found no evidence for higher cancer risk: "To date, there has been no persuasive evidence of any other health effect in the general population that can be attributed to radiation exposure."[5]

But even if we *were* to take this 4,000 death figure, nuclear would still have killed around 6,000 people over its entire history. That is tragic. But so is the fact that tens of thousands of people die prematurely *every day* from fossil fuels. We rarely think about that. A move from fossil fuels to nuclear power would add years of life expectancy for millions of people across the world.

Final note: For those interested in how the chart at the start of this question would look if we assumed that the Chernobyl death toll was higher (taking the 4,000 deaths estimate instead): it would move up above wind power, but still well below hydropower. This is why I don't recommend ranking wind, nuclear, and solar—because they depend too much on assumptions that we might never be able to pin down accurately. That they're far safer than fossil fuels, though, is not in doubt.

QUESTION 22

Doesn't it take too long to build a nuclear plant?

Answer: Nuclear plants in the West often have long delays, but some countries can build plants in six to eight years.

The US started pouring concrete for Vogtle 3 and 4—its latest nuclear reactors—in 2009. The units were expected to start producing power in 2016 and 2017. Instead, they came online in 2023 and 2024. That's at least 14 years to get a single nuclear unit connected to the grid. The UK has had a similar experience. Concrete was first poured for Hinkley Point C—its first nuclear plant in 20 years—in 2017. The first unit was expected to start producing power in 2027, but now it looks unlikely to happen before 2031. Flamanville in France started construction in 2007, with an expected completion of 2013. It finally opened in 2024: more than a decade late.

This is a terrible advertisement for nuclear power, especially at a time when we need to deploy clean energy quickly to get off fossil fuels. But it doesn't have to be this way. Some countries are building nuclear plants much faster, and countries like the UK were building them just as quickly in the past. I dug into construction time data on over 600 nuclear power plants built since 1950.[1] Here's what I found.

While there were some nuclear reactors with very long delays—which I'll come on to—most reactors (around 80%) took less than a decade to build. Some were built even quicker: One in five reactors took less than five years to build. So, the 14-year construction time for Hinkley Point C is not normal. Six to eight years is the average construction time globally.

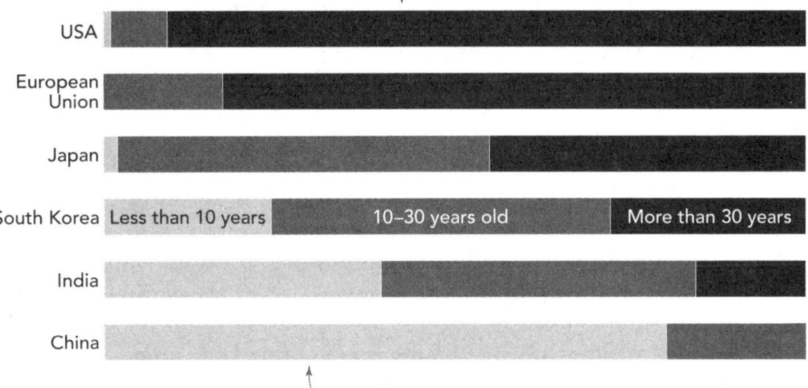

Figure 22.1
The age of nuclear plants around the world: the share of nuclear plants that are younger than 10, older than 30, or somewhere in between. Nuclear plants in Europe and the United States are old; those in China and India are young.
Data source: International Energy Agency.

But if we *used* to be able to build nuclear plants quickly—in the 1950s and '60s—are we just getting worse over time? At a global level, not really.[2]

What has changed is *where* power plants are being built. In the 1950s, '60s and '70s, it was all Europe and North America. France and Sweden were building loads of reactors, as was the United States. But most Western countries have essentially stopped building. Nearly all nuclear plants in Europe and North America are more than 30 years old.[3]

Plants in Asia are much younger. China and South Korea are building plants at a reasonable pace: in around five to seven years. India is building too, albeit more slowly. So, it's not that nuclear plants *have* to take over a decade.

Despite its struggles, nuclear power is not dead yet. It will grow in countries like China and South Korea, but not near the speed of solar and wind. If nuclear power is to play an increasing role in Europe and North America, it'll need to be built a whole lot faster.

It's worth reiterating that no country today is building nuclear as quickly as solar and wind. Even in very nuclear-friendly markets like China and

India, renewables are growing *much* faster. Nuclear power can be a great complement to these energy sources in building a stable grid system, but far more power is going to come from renewables simply because they can be built quicker and cheaper.

WHAT WE NEED TO DO

While nuclear might seem like it's dropping out of the clean energy race, there are things that would help to give it a fighting chance of a comeback.

- **If you start building nuclear plants, you need to keep going.** Countries are most successful at building nuclear plants when they have skilled workers who have experience from other recent projects. The problem is that countries like the US, UK, and France *had* a large skilled workforce but stopped building. Getting that expertise and experience back takes time, so a "one-off" nuclear plant is not going to be delivered quickly. Better to keep those skills fresh by building a series of plants.
- **Keep regulations for nuclear plants consistent.** One of the reasons why building nuclear plants in countries like the US has become so slow (and expensive) is because of constantly changing regulations. This means nuclear plants can't be built in the same way as previous ones. Of course, making reactors safe is top priority, but some might suggest that new regulations don't make them much safer and simply slow things down.
- **Standardize reactor designs.** Once you've found a great recipe, don't change it. Just build the same thing again and again.
- **Invest small amounts into innovation on "small modular reactors" (SMRs).** If one of the reasons for massive delays is that big nuclear plants are complex, then one solution is to make small ones that can be copied. These much smaller reactors could have a replicable design. SMRs are still young, so it's hard to know whether they'll take off. I'm a bit skeptical but think we should provide a bit of funding to find out. Companies like Google obviously think so; it has just ordered seven SMRs to power its data centers by 2030.[4] Amazon is looking into them too. Let's see if they're delivered on time.

THINGS TO BEAR IN MIND

All countries have had plants that opened well behind schedule. Why?

Nuclear plants are complex. There are many different types. The designs are complicated. And they are *big* projects. As Bent Flyvbjerg and Dan Gardner explain in their book *How Big Things Get Done*, this is a recipe for delays and cost overruns.[5] They studied a long list of construction projects—such as buildings, bridges, roads, and power plants—and nuclear came out worst for the risk of overruns.

The lack of modularity is its biggest problem. Compare the construction of a nuclear power plant to a solar farm. Almost every solar panel in the world is the same. It can be built quickly. You therefore iterate—and improve—the process over and over. Nuclear is the opposite. Almost every plant is different, and you often have multiyear gaps between building them. There is little room for learning how to iterate.

Another point is that "construction time" in figure 22.1 is measured as the time it takes from the actual building to begin—the first laying of concrete—to the date that commercial operation starts. It would be useful to look at the length of the planning stage as well, but I couldn't find any comparable data on this. With complex designs and delays on planning approvals, the construction time could be much longer.

QUESTION 23

Isn't nuclear power too expensive?

Answer: Nuclear power is expensive, especially in the US and Europe, but some countries are building it much cheaper.

The price tag is another big problem for nuclear power. Solar and wind prices have plunged over the last decade. Nuclear power has gone in the opposite direction. It has become more expensive in several countries, particularly the US, UK, Germany, France, and Japan.[1]

It hasn't always been this way. In the 1960s and '70s, there was a nuclear boom, with countries like the US, France, and Sweden rolling out nuclear plants at speed without breaking the bank. Not anymore. In the US, plants that started construction in the 1960s cost around $1,000 per kilowatt. The Vogtle 3 and 4 reactors that are being built today are probably going to cost at least $6,000 per kilowatt (yes, this is adjusted for inflation).[2] Why has it become so expensive?

Many projects run well past their deadline (see Question 22), and these construction delays don't come cheap. "Soft costs" or "indirect costs" make up most of the budget blowout.[3] That's largely about the cost of people: construction workers, engineers, design experts. "Indirect costs" make up the majority of cost overruns, with labor costs accounting for 80% of the total.[4]

Many people pin the blame on the increasing burden of regulation. And not just an increase in standards, but the constant *changes* in regulation I've already mentioned.[5] The problem with ever-changing regulations is that the design of nuclear reactors is never consistent. Sometimes a nuclear plant is already under construction, so the engineers have to rethink some parts of the design so that it's compliant with the new regulation. This can have major knock-on impacts across the entire project, making time and cost overruns inevitable.

As I've already mentioned, cost reductions come from building the same thing over and over. This has been almost impossible for nuclear plants. The United States has just over 50 plants; France has 18; and nearly all of them were built decades ago. The US Department of Energy estimates that you need to build at least 20 identical nuclear plants to really work out how much one costs.[6]

This might explain why costs have been falling in South Korea and China. They *have* been building nuclear plants regularly and might be taking advantage of some of the learning effects. Look at construction costs since 2000, and you'll find that prices in the US and the UK are four to five times

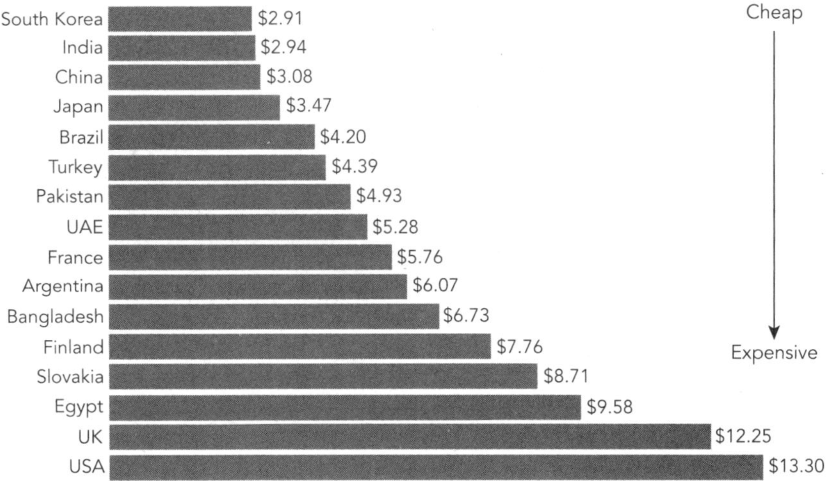

Figure 23.1

The cost of constructing a nuclear plant (in US dollars per megawatt). This is based on plants constructed since 2000 that have reliable cost data.

Data source: Sam Dumitriu and Ben Hopkinson (Britain Remade). Converted to dollars from pounds sterling using a conversion factor of 1.3.

higher than in South Korea.[7] This can't just be blamed on "cheap labor." Incomes in South Korea and Britain are pretty similar, so that doesn't explain a 500% difference in cost.

Where does that leave conventional nuclear plants today? Nuclear is probably not going to produce the bulk of electricity in most countries, and cost is the reason why. Per unit of electricity, it's more expensive than renewables, and often fossil fuels.[8] There are a few exceptions, such as South Korea, China, and India, where it can compete with coal and gas.

And yet once the reactors are built, they are relatively cheap to run, unlike fossil fuels where the cost of buying coal and gas to burn adds up. If you've already spent all that money on the nuclear plant, then keep it running and get your money's worth.

Nuclear could also compete on price in electricity grids that have high levels of solar and wind. In these systems, getting the first 70–80% of electricity could be very cheap. The final squeeze, though, could be more challenging. Some countries—like the UK—might need quite a bit of long-term storage to make it through winter (see Question 15). A combination of nuclear, solar, wind, and some storage might be cheaper than solar, wind, and storage alone.

WHAT WE NEED TO DO

Since unwieldy costs are often caused by construction delays, what we need to do is similar to my recommendations in the last question.

- **Standardize reactor designs.** Make a standardized reactor design that can be replicated and stick with it. Stop changing things.
- **Keep regulation and standards stable.** When I say "stop changing things" I mean regulation around nuclear plants too. You can't repeat a standardized design if it's no longer compatible with new regulations. Of course, this should never come at the cost of safety, so make sure you pick a standard design that is safe in the first place.
- **Maintain a skilled engineering workforce.** Train and invest in great people to lead these efforts, then hold on to them for dear life. Their skills and

experience will be invaluable if they can deliver a reactor on time and within budget.

- **Investing in small modular reactors is one way to get the benefits of rapid iteration.** Even with standardized designs, there are far fewer opportunities to build big nuclear plants than solar panels or wind turbines. This is why many investors and engineers are now drawn toward SMRs. See the previous question.

QUESTION 24

What about radioactive waste?

Answer: We know how to handle radioactive waste safely in deep geological sites, but countries need to prove it.

While I love *The Simpsons*, it has ruined the world's perception of nuclear waste. It isn't the neon green gloopy stuff you see on the TV. It doesn't seep out of metal canisters. It's not even a liquid. The fuel rods used in the reactors—which are the most radioactive and dangerous waste from the plant—are solid.[1]

Two important points before we dig into the specifics. The world's safety record on waste from nuclear power is extremely good: There have been zero deaths or injuries from it. Second, nuclear power is (quite rightly) the most regulated and tightly monitored waste industry in the world. The rules are strict, and there are no cutting corners. Much less attention is paid to waste from coal power, where the ash and residues from mining can be much more radioactive than waste from a well-regulated nuclear plant.

Nuclear power produces different types of waste. They have different levels of radioactivity, and that matters for how we treat them.

More than 90% of waste (based on volume) is low-level waste (LLW).[2] This is contaminated equipment from the operation of nuclear power plants, such as clothing, mops, tools, and protective shoe covers. Levels of radioactivity are low, so this waste can be stored on or near the surface in concrete vaults that are resistant to water, erosion, or tampering from humans. Most of the danger is gone within decades, but storage sites are currently set up to last 100 years or longer.

	Volume of waste	Radioactivity
Low-level waste (LLW) Gloves, shoe covers, etc.	92%	1%
Intermediate-level waste (ILW)	7%	4%
High-level waste (HLW) Used fuel rods	1%	95%

Figure 24.1

How radioactive is the world's nuclear waste? Most of the world's radioactive waste by volume is low-level waste with low levels of radioactivity. High-level waste, which is very radioactive, is produced in small quantities.

Data source: International Atomic Agency.

Intermediate-level waste (ILW) is more radioactive and has to be stored for longer. It's currently locked in multilayered casks (think of Russian dolls) at the surface, but many countries have proposed using storage sites underground.

That brings us to high-level waste (HLW). Less than 1% of the world's nuclear waste by volume is high-level, but it contains more than 95% of the radioactivity. It's mostly made up of the nuclear rods inside the reactor that run out of fuel—in the form of uranium—after a few years. Engineers have found a clever way to deal with this waste: reuse it. There is still unused uranium and plutonium in these rods at the end of their "life," which can be recycled and put back into the process as new fuel. Up to 96% of the material can be recovered and reused.[3] This not only reduces the amount of new uranium that needs to be mined, but also reduces the amount of waste that's produced.

The HLW that isn't reprocessed is initially stored in cooling ponds until levels of radioactivity and heat drop significantly, which happens in the first few years. It's then transferred into casks, which have multiple barriers isolating it from the environment, including a thick concrete layer. It's currently being stored at the surface but can't stay there forever.

How long this waste needs to be safely stored is a controversial topic. Some scientists estimate that after 10,000 years, the radioactivity levels will be lower than a natural uranium rock. Others advocate for longer periods: 100,000 or even a million years. But many experts expect that levels of direct radiation—what you would receive if you were close to the material—would be safe after 1,000 years, though it would still be dangerous if these materials were ingested via water supplies or food systems.

The solution that countries are banking on is deep geological storage: storing these multilayered casks hundreds of meters below the surface. This environment needs to be stable: Putting them in areas that are prone to earthquakes would be a terrible idea.

But, disappointingly, so far none of these sites have been built. While most nuclear scientists and engineers are confident that it *can* be done, we've yet to see one of the prototypes made in the real world. Finland is currently leading the way, with the Onkala repository due to open in the next few years. The site is 430 meters deep and was selected based on extensive testing of underground geological structures. It's designed to keep this waste safe for at least 100,000 years.

The amount of high-level waste that the world needs to store is relatively small. By 2016, the world had produced around 29,000 m^3 of the stuff. Nearly a decade has passed since then, so let's round it up to 40,000 m^3. If you were to put this into a football pitch, it would be just 5 meters tall. If that sounds like a lot, see how much waste is produced by fossil fuels every year in Question 17. Spoiler: It's a lot more than that.

WHAT WE NEED TO DO

Public perception of the nuclear waste problem is probably bigger than the nuclear waste problem itself. But there are a few things we can do to improve both at the same time.

- **Reprocess as much waste as we can into new nuclear fuel.** This is not only a smart way to deal with the waste; it also means we need to mine less uranium. Some countries—including France, Japan, Russia, and China—have policies and capacity to reprocess nuclear rods. The US doesn't yet have them, which means that they produce more high-level waste per unit of power, and miss out on a valuable source of energy. There is enough energy in US nuclear waste to power the country for a century.[4]

- **Hurry up and prove that we can store it safely underground.** Seriously, just get on with it. We can't keep saying that "it's possible to store nuclear waste safely underground" without doing it. Countries need to step up to show that it can be done and give a better sense of the cost (which could ultimately affect the affordability of nuclear power). Finland is hopefully going to be the first; others need to follow. Several countries have flirted with plans to build deep storage sites but have then struggled to get political and public support. The US has battled to get approval for the Yucca Mountain nuclear repository for decades without success. It'll be hard, but the more countries that achieve it, the more comfortable communities will be about storage sites. What most don't want to be are the guinea pigs that the rest of the world learns from.
- **Maintain strict standards for nuclear waste treatment and storage.** This is so obvious it hardly needs to be said. And it's already happening. But making sure that nuclear waste storage is taken seriously *everywhere* in the world is crucial. We're only one bad story away from public backlash that might be impossible to recover from.

V

ELECTRIC CARS

Transport accounts for around one-fifth of the world's CO_2 emissions, and three-quarters of that comes from cars and trucks on our roads. The most obvious way to get rid of road emissions is to walk, cycle, or take public transport. That's a no-brainer. We should be encouraging the development of more cycle-friendly cities and investing in public transport.

But, like it or not, the world is still going to have cars. If we want them to be low-carbon, our best option is to go electric. Yet the mainstream media is filled with an endless supply of misconceptions about electric cars: They're worse for the climate, terrible for air pollution, can't travel far enough, don't work in the cold. These headlines scare consumers back to the gas and diesel cars they're used to. None of these issues are deal-breakers, especially as the technologies behind electric cars get better and better.

QUESTION 25

Aren't electric cars just as bad for the climate as gas cars?

Answer: Manufacturing an electric car *does* produce more carbon than a gas one, but it quickly pays off this debt once you start driving it.

Less than half of Americans think that an electric car is better for the climate. I meet a lot of people who share that skepticism.

But the data is clear: Electric cars *are* better for the climate than gas and diesel. How much better depends on how clean a country's electricity grid is. You can see this in figure 25.1, which shows the average reductions in emissions from electric cars in different countries. In Europe, it's almost 70% less.[1] In the US, it's around two-thirds less (and better in states such as California). Even in China, with its coal-heavy electricity mix, you'd see a reduction of 37–45%. And in India, emissions are nearly a third lower.

Electric cars will get even better as we decarbonize our electricity grids. The UK government, for example, estimates that the average electric car currently emits 66% less than a gas car. By 2030 this will be 76% less.[2] These numbers are similar for the rest of Europe. In the US, life-cycle emissions would be around 70% lower (with even bigger reductions in some states with more clean electricity).

All these numbers already include the emissions from manufacturing the cars in the first place. "It takes a huge amount of energy to produce the battery in an electric car" is a common reason why people think they're so bad for the climate. And it's true that simply producing an electric car is worse for the climate than a gas one. So, electric cars start their life in a "carbon debt." But they quickly pay this back once they get on the road.

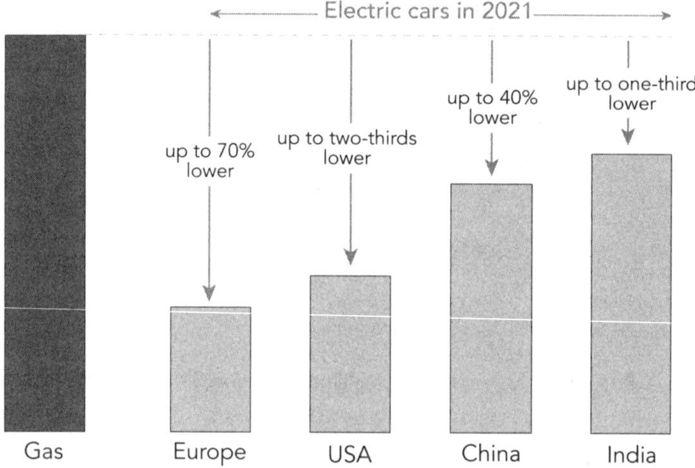

The carbon footprint of electric cars will be even lower in 2030...

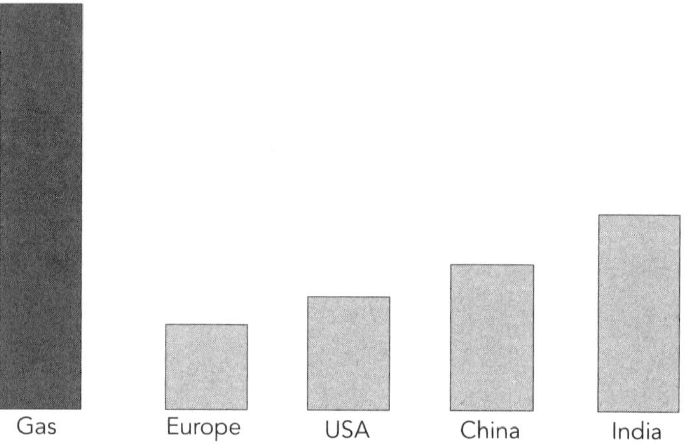

Figure 25.1

The carbon footprint of electric and gas cars: Emissions are measured over the lifetime of the car, including emissions from manufacturing and running it. This is measured in grams of CO_2 per kilometer.

Data source: Georg Bieker, International Council on Clean Transportation, 2021.

WHAT WE NEED TO DO

Here's how we can reduce emissions from road transport.

Walking and cycling still beat the car. That should be our default option if we can.

If you're buying a new car, opt for an electric one. It's better for the climate, and uses around one-third of the energy of a gas car. Buying an electric car is also a great signal to the market that this is what drivers want: Manufacturers get more confidence and produce better and cheaper electric cars.

Replace an existing gas car with an electric one. The climate benefits will depend on how much you drive. If you drive a lot—or your mileage is close to the average—this will be better for the climate. If you drive very little, the benefits are much smaller (or in some cases nonexistent).

Make the manufacturing process more efficient and cleaner. The emissions from producing the battery and the car itself have already fallen a lot over the last decade. CATL, the world's largest battery producer, recently slashed its manufacturing emissions by 45% in one year after rapidly increasing the amount of clean energy it uses.[3] If we can keep improving this process—and manufacture in countries with clean electricity—then electric cars will start their lives with a lot less carbon debt.

Decarbonize our electricity grids. The cleaner our electricity is, the cleaner our electric cars will be. There is a future where driving an electric car emits almost no greenhouse gases.

THINGS TO BEAR IN MIND

A common question I get asked is: How many miles do I need to drive before an electric car "breaks even" on emissions and is no longer paying off its carbon debt from manufacturing? It all depends on how clean your electricity is. The average driver in Europe needs to drive around 11,000 miles. In the UK, around 13,000 miles. This takes less than two years for the average UK driver.[4] In the US, it depends on what state you're in but this typically ranges from around 10,000 miles (in California, with a low-carbon electricity mix) to 25,000 miles in Missouri (which still runs on a lot of coal). That's a payback time of one to three years for the average driver.

Headlines in some newspapers will tell you that you must drive at least 55,000 miles before you break even. Or they'll cite a 2021 study from Volvo, which claimed that this number was 68,000 miles.[5] These numbers have been debunked many times. The corrected figure from the Volvo study is around 16,000 miles.[6]

The numbers I've used in this question are not cherry-picked from a single study. You will find the same results from many different organizations: the International Energy Agency, the International Council on Clean Transportation, the European Environment Agency, the UK and US governments, the World Bank's International Transport Forum, the MIT Trancik Lab, Carbon Brief, and a long list of academic papers from across the world.[7]

Some poorly researched headlines get such high numbers because:

1. They overestimate the emissions from electricity production.
2. They use very old, cherry-picked estimates for battery production.
3. And they underestimate how much gas-powered cars emit. They take the reported emissions figures of gasoline and diesel, but we know that in the real world the efficiency of cars is around 40% worse than it says "on the label." A final mistake is to forget to include the emissions from producing and refining the gas—this adds around 20% to the car's emissions.

Volvo admits that its numbers likely overestimate the emissions from electric cars. Yet the company is still very positive about its ability to tackle climate change, reporting that electric cars offer a "great reduction" in emissions compared to the internal combustion engine, and it has committed to going fully electric by 2030.[8]

So far, we've been comparing an electric car to a *new* gasoline or diesel car. But is it better to replace an existing one? The answer for most drivers is still yes. A new electric car will be in even more "carbon debt" when you first buy it, but the "payback" time is still short in most countries. In the US the payback time increases from around two years to four years. Unless you drive very little, it's probably better for the climate to replace your current gas car, if you can afford to.

QUESTION 26

Don't electric cars also contribute to air pollution?

Answer: Electric cars eliminate tailpipe emissions and reduce pollution, even if they're slightly heavier and have more tire wear.

Air pollution kills millions every year, and road transport plays a major role in this. Would moving to electric vehicles save lives? The answer is yes: Swapping gas or diesel for electric will reduce local air pollution, although it won't eliminate it completely.

To compare electric and gas cars we need to look at each of the different ways that vehicles create pollution. In the chart I've shown the four sources of emissions, and whether electric cars are better than gas cars for each.

The clear winner for EVs is the elimination of tailpipe emissions. Pollutants such as nitrogen oxides (called "NOx") and small particulates ("PM2.5") are generated from gas and diesel cars. NOx is terrible for our respiratory health and has been one of the biggest villains in our fight against air pollution. While modern gas and diesel cars emit much less than older models, it's impossible to eliminate NOx and PM2.5 completely. Electric cars have no tailpipe emissions. Since tailpipe emissions are the source of car pollution that has the greatest health impacts, this is a huge win for public health even if electric cars don't reduce any of the other sources of pollution (which they can).

Now, the cars might not emit pollutants directly, but surely the electricity plants powering them do? Researchers have studied this trade-off and found that electric cars still win when the emissions from power plants

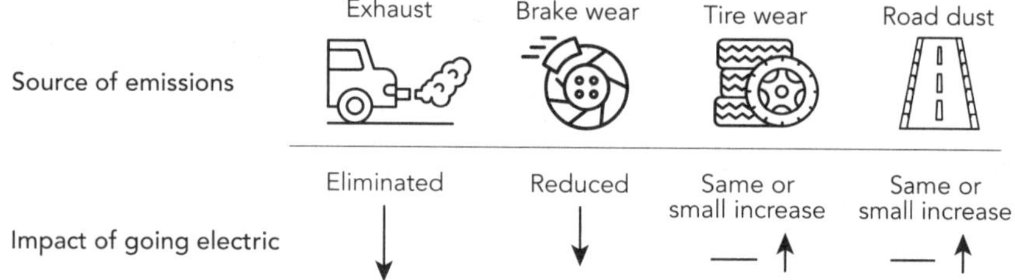

Figure 26.1
Electric cars and air pollution: Electric cars eliminate tailpipe emissions and reduce pollution from brake wear. Pollution from tire wear and road dust is equal to gas cars, or slightly higher if your electric vehicle is heavier.

are included.[1] This will only get better in the future as we decarbonize our electricity grids and stop burning fuels.

WHAT WE NEED TO DO

What can we do to reduce air pollution caused by electric cars even more?

- **Build cities around public transport, cycling, and walking.** This isn't rocket science: The most obvious way to cut car pollution is to not have one.
- **Lighter electric cars would help a lot.** Improvements in battery technologies and an emerging market for smaller, more affordable EVs could bring their weight down significantly. Gas cars are getting bigger and heavier too, with the rise of SUVs. If we're concerned about "car bloat" then a tax for gas and electric SUVs could help; France is one of the first countries to do this. As consumers, we're not off the hook: Opting for a smaller car over an SUV would help a lot. Your neighbors who need to squeeze into a parking spot next to you will probably be grateful too.
- **Use brake drums that capture particles from the brakes.** These drums can be placed on the brakes, where they stop particles from being emitted out into the road.[2] These only really work for electric cars, because the heat generated from inefficient braking in gas cars is too much for brake drums.

- **Innovate on low-wear tires.** There is still a lot of room for innovation on tires to produce models that wear down much more slowly.
- **Automated driving.** I'm sure you're a great driver, but many humans are not. If automated driving ever arrives, it will reduce the number of aggressive stops and starts, reducing the wear and tear on tires and brakes.

THINGS TO BEAR IN MIND

Electric cars eliminate tailpipe emissions, which is already a massive win for public health. But they can also reduce non-exhaust emissions. This is the combination of pollution from brake wear, tire wear, and road dust. How do electric cars affect each of these sources?

Brake wear: Cars produce small particles of pollution when they brake. The brake pads are pressed against the disks, and this causes them to wear over time. That wear and tear releases particles, which is a source of pollution. Brake wear is much lower in electric cars because they use regenerative braking: This converts the energy from the moving vehicle into electricity, which is then used to recharge the battery. Since resistance from the motor does a lot of the work of deceleration, there's less friction on the brake disks, and they wear down more slowly.

Tire wear: After many miles of driving, you'll need to replace your tires. Road friction wears them down, emitting particles along the way. There are a range of factors that affect the rate of tire wear, including the types of tires and driving style. Aggressive driving with high rates of acceleration, braking, and turning really wear down the tires. A key factor, though, is weight: Heavier cars wear down the tires more quickly. Many electric models are heavier than their gas equivalents because of the weight of the battery. I compared the weights of hundreds of cars and found that EVs were a few hundred kilograms heavier (although I think they'll get lighter over time as more affordable models enter the market).

The truth is that we don't have good studies that accurately compare the rates of tire wear in EVs compared to gas cars. Real-life anecdotal evidence suggests that rates of tire wear are pretty similar. The rear tires tend to wear at the same rate, and the front tires on EVs wear a bit faster due to increased torque—probably by around 10% to 15%.

Road dust: You're standing at the side of the road on a dry day. A big truck goes past, and you get dust blown in your eyes. Road dust is something that most of us experience but tend to forget when we think of "air pollution." In general, bigger vehicles resuspend more dust. However, this is probably more about aerodynamics (the *size* of the vehicle) than its weight. EVs that are a bit heavier or bigger than gas cars might increase pollution from road wear and dust a little.

QUESTION 27

Aren't electric cars too expensive for the average driver?

Answer: Electric cars are much cheaper to run, and will soon be just as cheap to buy upfront.

Cost has been the biggest barrier for electric cars.[1] But this is about to change because the cost of electric cars is tumbling.

To compare electric and gas cars, we need to consider two things: how much it costs to run, and how much to buy upfront.

The difference in running cost depends on the cost of electricity compared to gas, and how efficient your car is. In most countries, driving 100 miles in an electric car is now much cheaper than a gas one.[2] This is because electric cars use around one-third of the energy. *How much* cheaper depends on where you live, and the charging infrastructure available.

I did some back-of-the-envelope calculations for drivers in the US. Driving 100 miles in an electric car charged at home would cost around $4, compared to $13 in a gasoline one—around one-third of the price. The average driver in the US could save up to $1,200 per year on fuel. Or over $10,000 in 10 years. This is also true in the UK, where over the lifetime of the car, I could see more than £10,000 in savings.[3]

Electric cars win out almost everywhere. A study by Gulf Oil—yes, you read that right—surveyed the cost of running an average gas versus electric car in 33 countries and found that electric cars were cheaper to run in all of them.[4] Bad news for oil.

How you charge your car makes a big difference. As you can see in figure 27.1, public chargers tend to be more expensive. Standard public chargers are still usually cheaper than gasoline (although the costs can vary a lot, from $6 to $15 per 100 miles), but the fast-charging ones can be more expensive. There are exceptions to this: Tesla superchargers, for example, can get you 100 miles in $10 or less, depending on time and location. This is an obvious inequality that needs to be fixed: People living in flats without on-street or at-home charging (who often can't afford a house with off-street parking) could easily be priced out of electric cars because they're more expensive to run.

What state you're in also makes a big difference. I've used national average prices in these calculations, but the price advantage of at-home charging is much smaller in some states with expensive electricity. In California, it might cost you around $8 per 100 miles, or in Hawaii as much as $12. This is still usually cheaper than gasoline, but the savings are not as big as elsewhere.

Electric cars are also cheaper to maintain because they have fewer moving parts (such as valves and pistons in the engine), have less wear and tear on the brakes, don't require oil changes, and have no exhaust system or catalytic converters. Look under the bonnet of an electric car and you'll find a much simpler system. A recent report found that Teslas had much lower maintenance costs than popular gas cars on the market.[5]

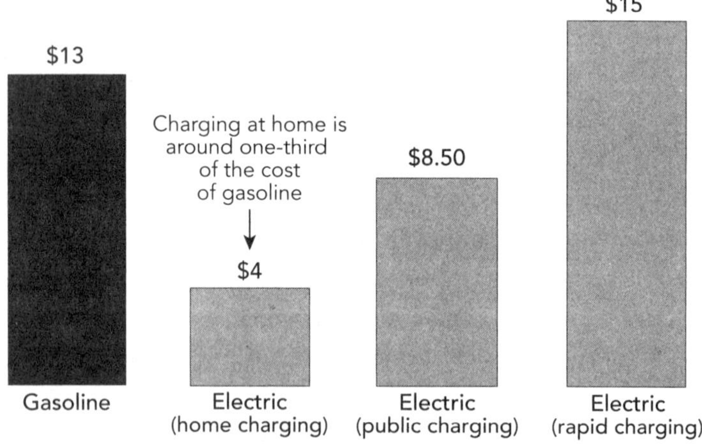

Figure 27.1
The cost of driving 100 miles in an electric versus a gasoline car. Costs will obviously vary depending on the efficiency of your car, and fuel prices where you live.

Lower running costs are nice, but not that useful if people can't afford an electric car in the first place. Aren't the upfront costs much higher than gas equivalents? Until recently, yes. Electric cars were for rich early adopters. But this is changing quickly. They're not quite there in every market—in the UK, most new electric car models are more expensive than their gas peers—but in many countries, EVs are approaching price parity.[6] It's staggering how quickly this gap has narrowed.[7]

One big reason is that new entrants are pushing into the car industry, especially from China. BYD, now the world's leading EV manufacturer, is selling well-designed, good cars in China for as little as $10,000. They have around 250 miles of range. Of course, they will get a massive price hike in the US and Europe to protect domestic car makers (see "What we need to do"), but these companies are proving that electric cars don't *have* to be expensive. And they're putting pressure on the well-established manufacturers to get cheap or die. This is exactly the competition the world needs if we want to move to low-carbon energy faster.

Electric cars have also reached price parity in *used* car markets, which is where most people get their cars from. I recently looked at used car prices in the UK and found that among the 10 most popular cars, the electric model was cheaper than the gas one for most of them. This was true for many of the bestselling brands: BMWs, Peugeots, Vauxhalls, Hyundais, Teslas, and more.[8]

Combine the running and upfront costs, and you find that the total cost of buying and owning an electric car is lower in most markets. As the UK's Climate Change Committee puts it: "Electric vehicles will be significantly cheaper than gas and diesel vehicles to own and operate over their lifetimes, so any undermining of their rollout will ultimately increase costs."

This rapid turnaround has been driven by massive innovations in battery design and the efficiency of manufacturing. But it has also been spurred on by more and more manufacturers pivoting from gas cars to electric ones. Car makers have realized that if they don't act now, they will be left behind. Ultimately, consumers are the winners as they get better EVs delivered at increasingly low prices.

WHAT WE NEED TO DO

What can we do to make electric cars even more affordable for the masses?

- **More electric cars sold today means cheaper electric cars tomorrow.** If you can afford to switch from a gas car to an electric one, do it. Batteries get cheaper the more we deploy. That means we're pulling down future prices, making them more affordable for consumers who might have gone with a new gas car instead.
- **Limit tariffs on imported electric cars.** There has been a mad rush to put tariffs on Chinese electric cars sold in Europe and the United States. This is bad for consumers as they are blocked from getting an affordable electric car. Western car makers are panicking: They've been complacent and are now far behind emerging competitors. Tough luck. This is how markets and competition work. Governments should limit these tariffs so consumers are free to decide for themselves.
- **Make electricity much cheaper than gas.** Whether an electric car is cheaper to drive depends on the cost difference between electricity and gas. Electricity prices in the US are relatively cheap compared to other parts of the world. They are not cheap in the UK, where most energy taxes are put on electricity whereas gas and gas get a relatively free ride. If energy taxes were distributed more fairly, the benefits of going electric would be even better. The UK *is* working on a plan to reform its electricity markets to make decarbonization more affordable.
- **Make sure that people have access to cheaper charging stations.** If someone has to rely on public chargers all the time, electric cars can be just as expensive as gas ones. Cheaper charging infrastructure needs to be available everywhere and we need to find ways of providing at-home charging to households who park on the street.

QUESTION 28

Aren't electric cars only good for shorter journeys?

Answer: Most low-range electric cars now get over 200 miles, and many models are getting well over 400 miles.

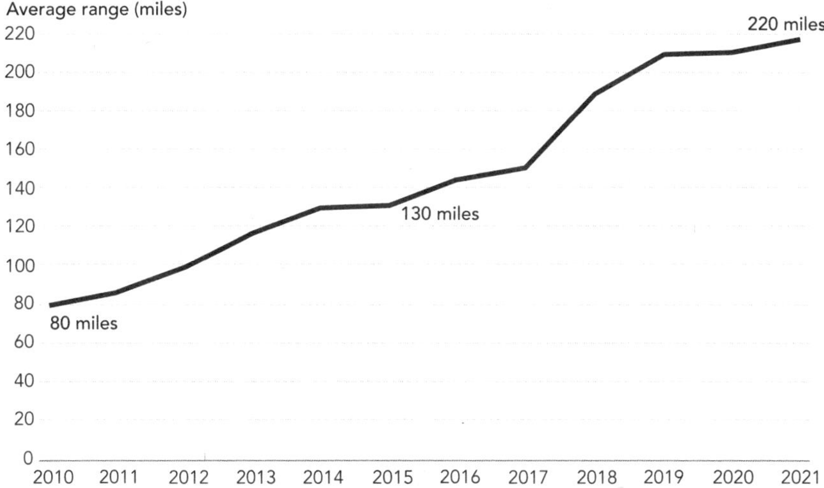

Figure 28.1
The increase in the average range of electric cars. The average electric car can now travel more than 220 miles, and some can go over 450 miles.
Data source: International Energy Agency.

In 2010, the average electric car could travel just 80 miles on a single charge. By 2021, this had nearly tripled to 220 miles.[1] That's the global average. The United States, which prides itself on bigger and more expensive cars, is reaching an average of 280 miles. There are now a bunch of models with ranges of 450 miles or more.

It's not just the average range that has increased a lot; manufacturers are pushing the upper bounds of what range is possible. In 2011, the car with the longest range was the BMW ActiveE, which would give you 94 miles.[2] Then, for years, Tesla took the top spot. But in 2022 it was beaten by the luxury Lucid Air Dream R, which manages a whopping 520 miles. This record will probably be beaten by the time this book reaches you: The industry is moving so fast it's impossible to keep up.

What's still a problem is "charger anxiety": whether there are enough working chargers along your route. "Range" and "charger" anxiety might seem like the same thing, but they're different. Imagine the same example for a gas car that has a 250-mile fuel tank. If there were no gas stations on your 250-mile route, you wouldn't blame the car: You'd blame the government for skimping on gas stations. The same is true for electric cars. If there are no working chargers on your 250-mile road trip, the problem is the infrastructure, not the car. That's the question we're going to look at next.

WHAT WE NEED TO DO

How do we ease people's anxieties about car range?

- **Give drivers what they want, not what we think they need.** If drivers want the option of choosing a more expensive 500-mile range car, then give it to them. Don't try to convince them that a 200-mile range car is all they need. This is only going to create another unnecessary block for people who would otherwise be willing to go electric.
- **Make sure there is good public charging infrastructure.** Having chargers is not enough: They also need to work and not have 20 drivers queuing for them.
- **Share positive experiences if you have an electric car.** Many people have nightmares about running out of charge on the motorway. But

this is rare. If you have an electric car, sharing good news and being honest about how to navigate longer journeys will help ease people's concerns.

- **Consider renting a gas car for your once-a-year road trip.** If that one long journey you take every year is a deal breaker, then one option is to get the electric car and just rent a gas one for that single trip. This *should* be totally unnecessary (and if everyone was doing this a decade from now, I'd consider EVs to be a failure) but if it makes you feel more comfortable about long journeys then it's worth considering.

THINGS TO BEAR IN MIND

The range of electric cars has grown rapidly, but it's still worth asking whether this range is "enough." The average range in 2022—250 miles—is much further than most people travel in a day. In the UK, the average driver goes 20 miles. This is even less if the family owns a second car. The average American drives twice as far: around 40 miles per day. Still nothing to worry about, even with the cheapest and lowest-range electric cars.

The main concern, then, is those rarer long-range trips: traveling across the country for a holiday, or to see distant family and friends. The first thing to note is that the driver's "battery" will usually need a top-up before an electric car does. Let's take an extreme example where you're driving on the motorway at the speed limit. In four hours, you'd drive 280 miles; just a bit more than the average 250-mile electric car. Drivers should take a 15-minute break after a few hours of driving, so after four hours, it's time for them to stop for a rest. Perfectly timed for when the car needs a top-up charge. As long as there is a charger available, having an electric car doesn't cause much bother. In the future, many people will have an electric car that lasts 350 or 400 miles, so for them the journey will be even easier.

There are also concerns that the battery in an electric car degrades quickly over time, so that within a few years, your range is far lower than it was at the start. While batteries do degrade over time, these stories are often exaggerated. Studies across tens of thousands of vehicles have found that they degrade at a rate of around 2% per year. Some models degrade more slowly, at a rate of around 1% per year. So, if you've got a car with a 300-mile range, it might lose up to 30 miles after five years.

Most car manufacturers offer at least eight years of warranty on the battery. Even by then, it should still be going strong. The quality of batteries is going to keep improving over time, so rates of degradation will keep dropping and their lifetime will keep increasing.

There are things you can do to maintain the health of the battery. Keeping it between 20% and 80% of charge helps. Avoid large temperature swings. Park in the shade on hot days. And don't use rapid charging often: only when you need it.

QUESTION 29

Aren't there too few charging points?

Answer: Some countries are managing to keep up with the rise in demand for public chargers, but many are falling behind.

As we've just seen, what most people call "range anxiety" is more likely to be "charger anxiety." This is a very real and genuine concern. Forty-one percent of drivers say that they would only consider switching to an electric car if the availability of chargers was equal to that of gas stations.[1] Of course, this is a bit of a chicken-and-egg problem, but one that ultimately leads to a positive cycle if governments and companies are willing to invest. Charging points are installed first; people buy electric cars in response; this increases the pressure to install more chargers; and more people buy cars in response. The availability of charging points is a good predictor of EV adoption.[2]

More than a million public charging points are being installed across the world every year. The EU's charging network tripled in just three years, and it's already well ahead of its deployment targets.[3] And both public and private investments in charging infrastructure are increasing. The key question is whether there are "enough."

In 2023, the world had around 10 electric cars per public charging point, but this varied from 88 cars per charger in New Zealand to three in Korea. Does this mean that countries with high numbers of cars per charger are falling far behind? Often, but not always. This ratio is a useful guide, but not perfect. It misses two important factors.

The first is the *speed* of charging. You can get away with fewer chargers if they charge quickly. So, the EU has another guide, which adjusts for

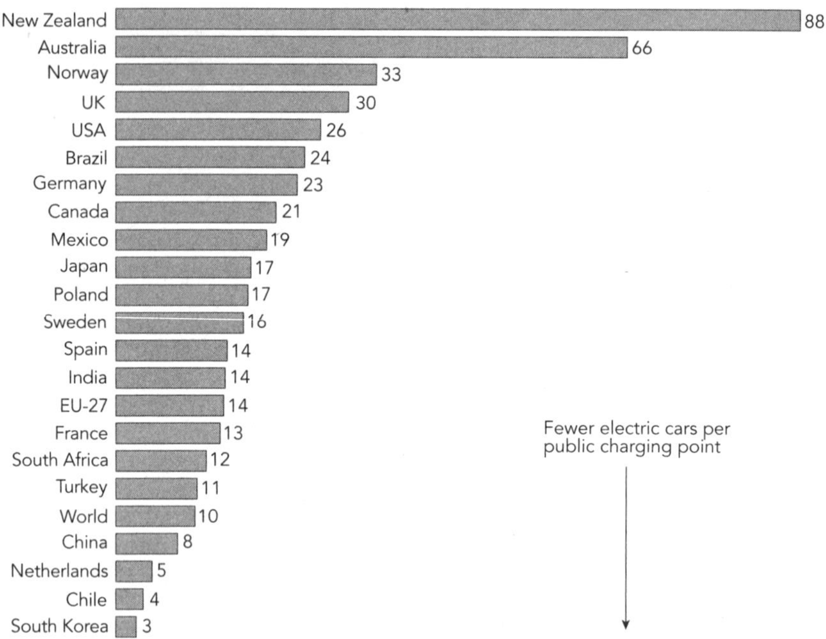

Figure 29.1

The number of electric cars per public charging point. Higher numbers indicate more electric cars to share public charging points.

Data source: Author calculations based on data from the International Energy Agency.

the power of each charger: 1 kW of public charging per electric vehicle. Again, some countries are meeting their targets on this, but a lot are behind the curve.

The second is that the demand for public chargers is not the same everywhere. Most people charge their car at home. This is especially true in countries like Norway and the US. Only around 5% of Americans use public chargers as their main source compared to 20% in the UK.[4] Dense cities make public charging harder: 70% of homes in the UK have garages or off-street parking, but just 30% in cities do. People also charge their cars at work: In the US, 30% of EV drivers almost exclusively charge at their workplace.[5]

Most drivers don't think that public networks are managing to keep up. In a survey of 31,000 EV drivers in 15 countries, 70% said they don't think there are enough.[6] Eighty percent of people considering an electric car think so too. Even if some of this is just a gap in perception, it's not an issue we should ignore. It will stop some people from making the switch. But interest and deployment in charging networks are increasing, from governments and the private sector. Even fossil fuel companies are jumping on the bandwagon, with commitments from companies such as Shell to install charging points in their gas stations. This in itself is an obvious indicator of where the world is heading: Road transport is going electric, and oil companies know it.

This—next to cost—is one of the biggest barriers to EV adoption. Some countries are on track and keeping up with demand, but others are lagging. The fact that some are doing well means it's a tractable problem.

WHAT WE NEED TO DO

How do we make sure that public chargers are available for those who need them?

Cities and remote areas should be first in line. The people who need public chargers the most are those who live in dense cities or those in rural areas where transport infrastructure is scarce. This is where efforts should be focused—not suburbs where most people can charge from home.

Invest in charger maintenance, not just installation. It's easy to focus purely on the number of chargers that have been installed. But making sure they're consistently in good working condition is critical to maintaining trust among drivers.

Governments need to work with private companies to build out the network. A lot of these efforts can be publicly financed, but there's also a good business model for private companies to supply and build out charging networks. Tesla has already done this in many countries and made it part of its business portfolio. Since company reputations are on the line, there's pressure to keep these chargers running smoothly.

QUESTION 30

Won't electric car charging break our electricity grids?

Answer: With smart charging and vehicle-to-grid systems, electric cars can be an asset rather than a liability to the grid.

There are around 250 million cars on the road in the US. Imagine if they were all electric and everyone came home in the evening and put them on charge at the same time. Peak power demand in the US for cars alone would be at almost 2,000 gigawatts (GW). Peak power demand in the US today is just under 1,000 GW, so the country would need at least twice as much *just* for its cars. The electricity grid would collapse. This is the silly math people use to show that our electricity grids can't cope. But these numbers don't make sense. Yes, if everyone tried to charge their car at 6:30 every night, this would be a problem. But they won't.

First, taking the US average of 40 miles per day, and the UK of 20 miles, most people won't need to charge their car every day. Most drivers could charge their car once a week and they'd be fine.

Second, there will certainly be an evening after-work peak, but some drivers might charge during the day, others in the morning, and there would be a spread of charges at different times in the evening. In Norway, 25% of cars on the road are already electric, and its grid is coping fine.

Third, we can design a smart grid system that *makes sure* the cars don't all charge at the same time. This system—called "vehicle-to-grid" (V2G)—would turn electric cars into an *asset* rather than a liability, so EVs can act as batteries for the grid. Your car charger can "speak" to the grid and charge when our electricity supply is high and prices are cheap. Then sell power back to the grid when demand is high and prices are expensive. Consumers might even make a bit of profit. In windier states, for example, the smart grid might charge your car during the night when overall demand is low and there's lots of excess wind power. In one with lots of solar power, it might charge it at midday when there's lots of sun that might go to waste.

Of course, you want to make sure your battery is charged for when you need it, but the system can be programmed to know when you'll be going out. And it would never leave your battery completely empty in case you suddenly need to make a trip.

When people talk about EVs breaking the grid, they're usually talking about *peak* demand for power. I ran the numbers to see what impact electrification would have on our electricity demand overall.

If all of the UK's cars went electric tomorrow, electricity demand across the year would increase by around 20% to 25%.[1] In the United States, EVs would increase electricity demand by a similar amount: around 25% to 29%.[2] With all the trucks and vans included, this would be 33% to 50% higher.

So, we should expect a large increase in electricity demand from going all electric on transport. It could mean increasing the size of power grids by 50%—from 4,300 to as much as 6,500 TWh, which is an expansion the US hasn't done for long time. But if countries are serious about getting to net zero, they know they will need to double or even triple the amount of electricity they generate; road transport is a major driver of this demand, but far from being the only one.

What's more, national grids across the world see electrified transport as an ideal solution for balancing the grid, rather than a threat that could break it. It could be a vital part of the solution for the peaks and dips in the electricity system.[3] Studies suggest that V2G solutions can help to reduce price volatility, reduce consumer prices, and lower the overall operational costs of a charging system.[4]

If we use batteries smartly, they make our grids more resilient. If there's a disaster—like an earthquake, a major storm, or a freak cold snap like Texas has endured in recent years, then it might be these batteries that keep us running, for a few hours at least. A move to electric cars will

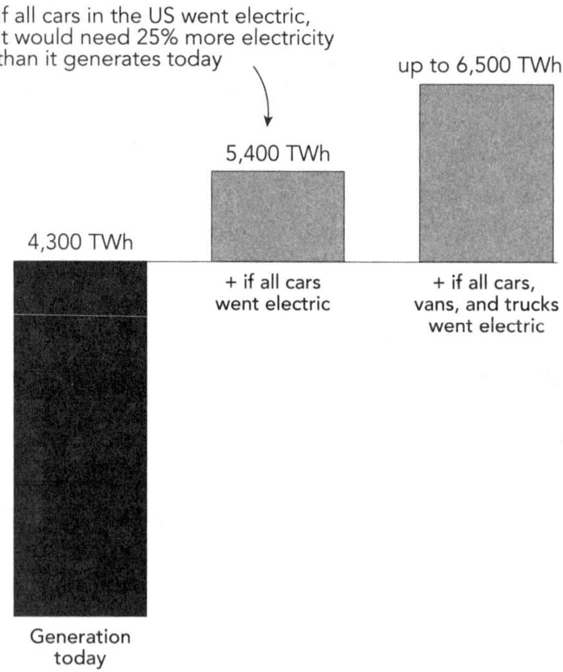

Figure 30.1
How much extra electricity would be needed if all cars in the US went electric, and if all vans and trucks did too? This is compared to current electricity production.
Data sources: Ember Energy, US EIA, and author calculations.

increase demand for electricity, but not as much as many people claim. And if we make this transition properly, it will help our grids rather than tear them down.

WHAT WE NEED TO DO

If you own an electric car, there are two things you can do to help. The first, as always, is to simply drive less: walk, cycle, take public transport if you can. Less charging means reduced strain on the grid. The second is to charge at times when electricity demand is low or supplies are high: during the night, or at midday if you live in a sunny country with lots of solar power. This not only helps the grid but could make it cheaper for you to charge too.

There are also a few things we can do to incentivize other drivers not to charge their cars at "peak times."

- **Smart metering to optimize cheap prices.** Even without fancy V2G technologies, having smart metering that tells users (or the car) when electricity demand is low and prices are cheapest will help spread out charging times across the day, and away from the peaks. Smart metering is already quite common—and many people take advantage of it—but making it more obvious and easier for consumers would help a lot.
- **Trial and scale up vehicle-to-grid networks.** Despite their potential benefits, no country has rolled them at large. They have been trialed at much smaller levels, but to reap the benefits, these systems need to be ramped up quickly. This will need investment and innovation from private companies—Octopus Energy in the UK is just one example of a company already working on it—and government legislation that makes it possible to scale. The UK, French, German, Dutch, and Swiss governments are already leading the way on making this legal. In the US, there is no national or state-wide rollout of V2G, but there have been a range of smaller-scale trials stretching from New York to Colorado. California is arguably the furthest ahead; working with the large utility Pacific Gas and Electric (PG&E), it has launched a number of residential programs and a fully electric school bus system that integrates vehicle-to-grid capabilities.

QUESTION 31

Don't electric cars struggle in the cold?

Answer: Electric cars do lose around 10% to 20% of range in the cold, but they are still thriving in some of the world's coldest climates.

It's true. Electric cars lose some of their performance in the cold: The range of the car is reduced, and it takes longer to charge. But the claim that electric cars "don't work in the cold" is blatantly false.

The Norwegian Automobile Federation (NAF) has done the most extensive tests on 50 different EVs in summer and winter conditions.[1] In its winter 2022 test, the cars were taken on a standardized driving route, with the temperature varying between 0 and –10°C. They drove each car until the battery ran out and compared the range to the advertised one. Every car had a lower range than what was advertised. But the amount varied by model. You can see this in the chart.

The average drop was 19%. So, if your car had a labeled range of 450 kilometers (280 miles), you might only get 364 kilometers (226 miles) in winter. The best-performing cars were the BYD Tang and the Tesla Model Y: They lost just 11% of their range. At the other end, the Skoda Enyaq iV80 lost a whopping 32%, a result that was so controversial the Norwegians had to test it again with another car (they got the same result, so it wasn't a fluke).

These results are consistent with results I've seen elsewhere. Most studies report a 10% to 20% drop in range in cold temperatures, with a few cars losing a third.

This drop in range will not be a deal-breaker for most drivers. The most popular EV models are still getting at least 300 to 400 kilometers, even in

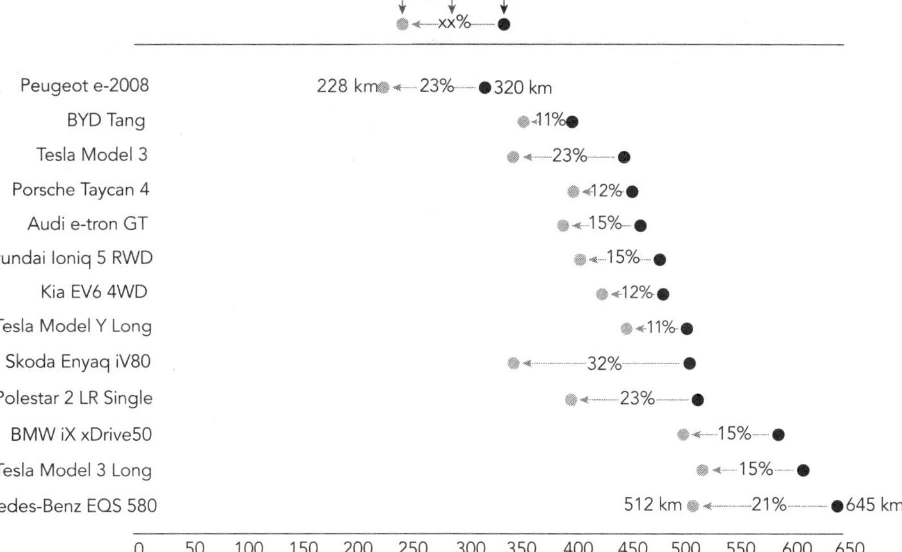

Figure 31.1

Loss of range in cold temperatures: the reduction in electric car range for a range of models when driven in cold temperatures. Based on standardized tests in Norway.

Data source: Norwegian Automobile Federation.

cold temperatures. Most car journeys are under 10 kilometers. In the US, just 5% are longer than 50 kilometers, so the percentage that gets anywhere close to 300 kilometers is tiny.

People also worry about *how long* it takes to charge in the cold. The NAF tested this too. With a fast-charger, going from 9% to 80% of charge took an average of 37 minutes in winter. That compares to 33 minutes in the summer. So, winter charging was 10% slower. Not too bad. There were a few cars—such as the BMW iX—that took quite a bit longer: Charge time increased by around a third. Some models, though, managed to charge just as fast in subzero temperatures.

Perhaps the strongest evidence that electric cars can tolerate cold temperatures is that the countries where they're most popular are, well, pretty cold. In Norway, almost every new car sold is electric. For four or five months of the year, temperatures drop below zero. In 2023, the other countries that completed the "most popular" top five were Iceland, Sweden, Denmark, and Finland. The high-latitude Scandinavians are EV enthusiasts despite cold, harsh winters, so the cold can't be such a big deal.

One final point: Gas and diesel cars are also less efficient in the cold and get fewer miles per tank of gas; really, no different from an electric car.

WHAT WE NEED TO DO

While the drop in range due to cold temperatures isn't a deal-breaker for most drivers, there are things we can do reduce the impact and make things more transparent for consumers.

- **Install heat pumps in electric cars.** Manufacturers are working to improve battery performance, and range loss will probably become even less of an issue over time. Some have already installed heat pumps in new models to keep the batteries warmer.
- **Preheat the battery before charging.** If you want to speed up the charging time of your car, preheating the battery should help. Many cars now have this feature.
- **Manufacturers should provide "cold range" numbers on their labels.** The reduction in range in cold temperatures might not be a deal-breaker, but drivers have the right to transparent numbers on what range they should expect in the winter.

QUESTION 32

Don't electric cars catch fire all the time?

Answer: Electric cars catch fire *less* often than gas cars, but can be harder to put out.

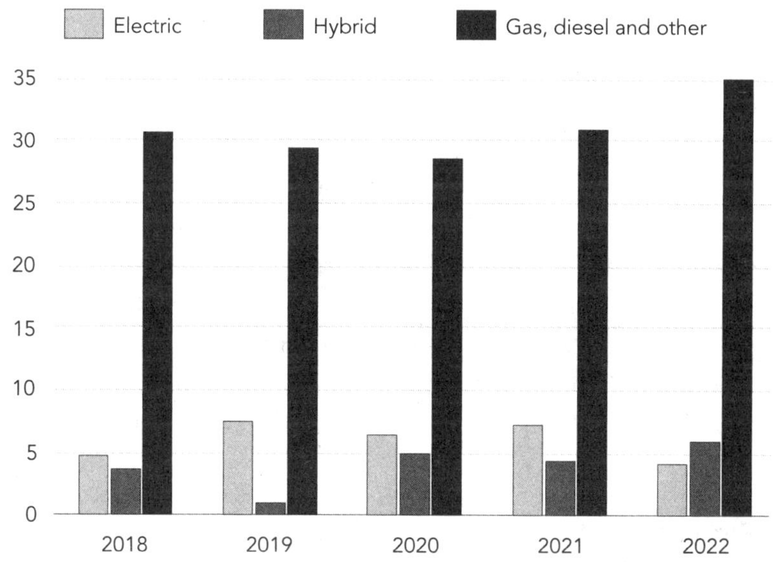

Figure 32.1

How often each kind of car catches fire. Measured as the number of passenger cars that caught fire per 100,000 cars of that type. Based on data from Norway.

Data source: DSB/SSB via Robbie Andrew, CICERO.

You might have seen scary headlines about electric cars "spontaneously catching fire."

Based on these (often wrong) headlines, a lot of people think that electric cars catch fire more often than gas ones. But this isn't the case. Fires in electric cars seem to be much *less* frequent. The reason I say "seem to be" is that very few countries have gathered detailed statistics that would let us compare fire rates over a long period of time. Here I'll draw on data from two countries where electric cars are now very common: Norway and Sweden.

The Directorate for Civil Protection in Norway looked at how many passenger car fires were reported in three different car types: electric cars, hybrids, and others (which is mostly gas or diesel). This is shown in figure 32.1. Gas and diesel cars were about nine times more likely to catch fire than electric ones. Data from Sweden found similar results: 0.004% of electric and plug-in hybrid cars were involved in a fire.[1] For gas cars, it was 0.08%—19 times as many. (One factor that might play a role here—as I'll discuss soon—is that the electric car fleet is generally younger than the gas cars on our roads.)

Just to put the Norway numbers in perspective: There were around four fires for every 100,000 electric cars on the road. That means 0.004% of EVs caught fire in 2022. Electric car fires are extremely rare.

EV FireSafe—an organization supported by the Australian Department of Defence—studies EV battery fires and provides training for first responders on how to deal with them.[2] It found that, based on global data from 2010 to 2020, there was a 0.0012% risk that your electric car catches fire.

WHAT WE NEED TO DO

While electric car fires are very rare, they are a dangerous hazard and there are things we can do to reduce this risk.

- **Train emergency services to handle electric car fires.** Electric cars are still relatively new, and many emergency services staff won't be trained on how to handle these fires when they do break out. They'll be trained on gas cars. Providing training—which is what organizations like EV Firesafe do—specific to EVs would reduce the risk of serious injuries.

- **Improve battery safety.** Car manufacturers like BYD (from China) have far safer batteries than other brands. They do heavy testing on their batteries, deliberately puncturing them and measuring the increase in the temperature. They rarely overheat. Hopefully, battery technologies will continue to improve, and this will be standard in most electric cars in the future.
- **News reporters need to stop assuming every car fire is an electric one.** Wait until you have the facts before assuming—and falsely reporting—that the latest fire was caused by an electric car. In 2024 there was a major fire in the parking lot at one of London's airports. Headlines shouted that it was an electric car that started it. It turned out to be a diesel Range Rover. There are countless other examples of similarly sloppy journalism.

THINGS TO BEAR IN MIND

There are two important caveats. The first is that the data on fire rates doesn't provide breakdowns by the *age* of the car. Most electric cars are relatively new to the roads, so the average age of EVs will be lower than gas cars. Since we might expect older cars to catch fire more often, this could partly explain why gas fires are more common. To do these comparisons fairly, we'd want to compare young electric cars with young gas ones and do the same comparison for old cars. Unfortunately, we don't have age-specific data to do this. We will need to wait for electric cars to age, so we can test whether they get more flammable over time.

Second, the *frequency* of fires doesn't tell us anything about how hard they are to put out. Despite what many headlines suggest, electric cars don't just spontaneously combust. But, like a gas car, they might catch fire if they're involved in a collision. This can cause the energy in a lithium-ion to short-circuit, causing it to heat up rapidly. Putting out an EV fire *is* more challenging than a gas one. They tend to burn hotter, so suppressing the fire takes longer and needs more resources. And the challenges overall are different: That's why companies such as EV FireSafe are focused on building training resources.

To be clear, emergency services *can* handle EV fires, but a fire in an EV might be more dangerous than one in a gas car. What the data makes clear, though, is that fires in electric cars are *very* rare. There are no signs that we're going to have parking lots full of EVs spontaneously catching fire.

VI

MINERALS

A world that runs on renewable energy is going to need a much wider range of minerals and materials than the world we've been running on fossil fuels. We'll need to dig more lithium, copper, cobalt, silver, graphite, platinum—the list goes on. That raises a lot of red flags. Won't we run out of minerals? What about all the mining that's required—surely that can't be clean or sustainable? The terrible environmental damage and the social injustices in supply chains? And what about the huge energy security threat from relying on a handful of countries to source and refine the critical elements needed to keep the world ticking over?

People are right to be concerned about the dynamics of a new global economy, and what risks it might create. But what's often ignored are the damages, injustices, and energy security threats that our current system—one mostly run on fossil fuels—already creates. Critical minerals for clean energy present us with a different set of challenges. But these are not enough to stand in the way of moving away from fossil fuels.

QUESTION 33

Won't the world run out of minerals?

Answer: We already know we have enough of most minerals, and we keep finding more as we go looking for them.

Humans are programmed to be on alert for scarcity, to think we're on the brink of running out of stuff. Water, food, even fossil fuels. We have, seemingly, been on the brink of running out of oil for decades. The new fear, then, is that we will run out of minerals to power clean energy. This seems unlikely for several reasons.

We tend to respond to scarcity by finding more stuff. Look at figure 33.1, which shows how much lithium—a key mineral in most batteries—the world produces every year, and the amount of reserves (deposits in the ground that we can extract) we know we have. Our reserves are going up even though we're using more: We're finding lithium deposits much faster than we're consuming them.

Manufacturers can also substitute one mineral for another when supplies are looking tight. When copper prices are high, aluminum can be used for power grid lines instead. When cobalt supplies are strained, some electric car companies—including Tesla—switched to battery chemistries that didn't use any cobalt at all.

Combine this with improvements in recycling and efficiency (see "What we need to do"), and, in the future, we'll probably need fewer minerals than we expect and will have more supplies to use.

Here are some sound bites from the (many) organizations, researchers, and industry experts who agree that long-term mineral demand is not a key concern.[1]

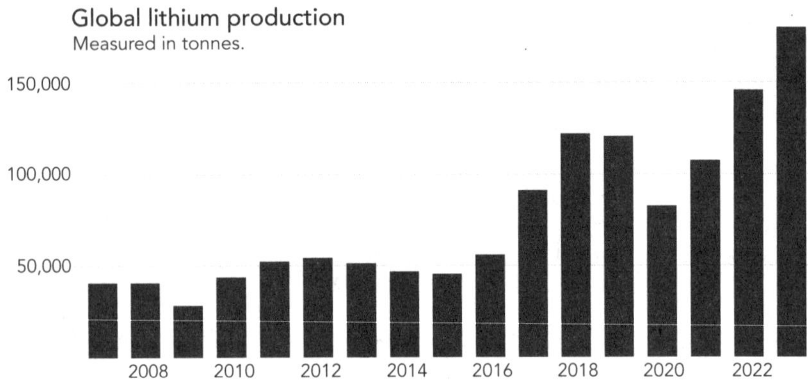

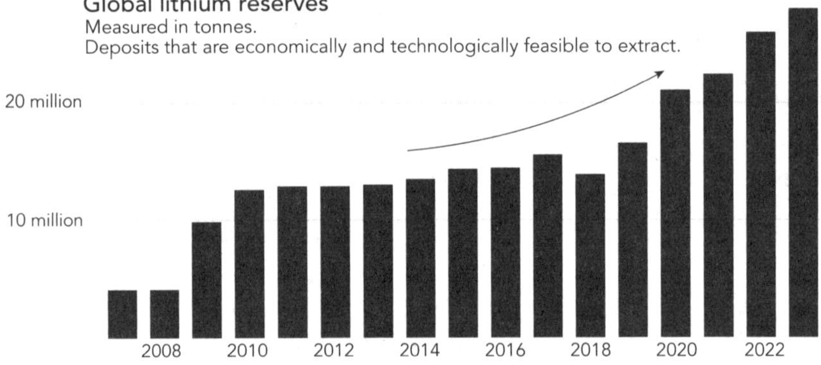

Figure 33.1

How fast we're finding lithium and how fast we're using it: The world digs more lithium out of the ground every year, but known reserves are growing far faster. When materials are in high demand, humans get better at finding them.
Data source: US Geological Survey.

- **International Energy Agency (IEA):** "There are generally no signs of shortages in the amount of available resources."
- **International Renewable Energy Agency (IRENA):** "There is no scarcity of reserves for energy transition minerals. . . . Production has surged for many energy transition minerals, and reserves mined from economically viable sources have grown."
- **Energy Transitions Commission:** "There is no fundamental shortage of any of the raw materials to support a global transition to a net-zero ecaonomy."

- **Rocky Mountain Institute:** "The world has enough minerals for the entire energy transition, and is rapidly building out the mines to extract them."

We can also look at specific studies that crunch the numbers to calm our fears about running out. We can compare *demand* for minerals in a net-zero world, and how much of each mineral there is in the ground.* Of course, we don't know *exactly* how many solar panels or electric cars the world will have in 2050, but we researchers can run a range of scenarios where the world looks slightly different in each.

For most minerals, global reserves are already higher than what we'll need. Clean energy won't have a big impact on the demand for these minerals; our wind turbines, batteries, and solar panels would only increase global demand for many materials by around 5–15%.[2] Supplies will be tighter for a few materials, such as lithium, cobalt, and copper. The energy transition will drive large increases in the production of these minerals, but we know that the world either already has enough or will manage to close the gap by finding more deposits, substituting for alternatives and becoming more efficient.

WHAT WE NEED TO DO

While it seems unlikely that running out of minerals is going to stand in the way of the energy transition, there are ways to reduce the risks of supply shortages and minimize the amount of materials we need to pull out of the ground.

- **Ramp up recycling rates.** More recycling means we need to dig less stuff out of the ground. It moves our energy system from being continually extractive to being more circular. One of the problems with recycling is that it's often *more* expensive than digging new minerals out of the ground. But the economics can tilt when minerals are in high demand

* There are two important terms that analysts use here: *reserves* and *resources*. Reserves are deposits that we can extract economically with current technologies. Resources are known deposits that also include those that *aren't* economically viable to extract under current conditions. Reserves and resources can both increase over time as we discover more deposits and others become economically viable.

and get expensive. High prices encourage manufacturers to invest in recycling, which eases pressure on mining. That's better for the environment too, so it's a win–win.

- **Improve material efficiency.** Building a solar panel or battery today uses fewer materials than a decade ago. Jenny Chase, possibly the world's leading expert on solar markets, estimates that the amount of silver used per solar panel has halved in just five years. That means the silver from a five-year-old solar panel would provide enough for two panels today. Solar panels need just one-eighth of the polysilicon that they did in 2004. Panels now coming toward the end of their life would have enough polysilicon to produce eight panels today. We need to focus on improving material efficiency so that this keeps getting better.
- **Invest in mining projects early and ramp up refining capacity.** While people fret over a long-term problem that doesn't exist, they're not focused on the real bottleneck: opening mines and building refining capacity *now*. While long-term supplies will be fine, they could be tight in the next 5 to 10 years as electric car and solar panel markets pick up speed. It's not about whether there are enough minerals in the ground; it's whether we can get them out in time. The discovery and exploration phase for finding minerals can take over a decade. Some countries can do it faster: Lithium sourcing in Australia needs around four years. It's six years in South America. But even once a deposit has been found, it still takes around five years of permitting and construction before that mine starts producing. Some countries are even slower. In the United States, it can take 7 to 10 years to secure a mine permit.[3] That's too slow. We need to hurry up.

QUESTION 34

Won't renewables and electric cars mean a lot more mining?

Answer: Shifting from fossil fuels to clean energy will reduce demand for mining, not increase it.

To move to low-carbon electricity and electric vehicles, we'll need to mine a lot of copper, nickel, lithium, manganese, and other minerals. The road to clean energy involves a lot of digging and building; there's no point in pretending that this won't leave any footprint.

But it's wrong to think that mining will explode with the shift to clean energy. We'll need to dig *less* stuff out of the ground in a low-carbon energy system than we currently do with fossil fuels. Figure 34.1 shows the mining footprint of different technologies that are used to generate electricity.[1] The numbers include all the materials from construction and manufacturing through to the fuel that's burned in the power plant itself. But they also include all the ores and rocks that need to be moved or extracted to get those minerals out.

As you can see, solar, wind, and nuclear power require much less mining than coal. Coal power needs around 30 times more mining than renewables, more than 100 times more than nuclear. Gas also needs more extraction than low-carbon sources, although there is a caveat to this (see "Things we need to bear in mind").

What people often miss when thinking about these comparisons is the fossil fuel itself. Sure, it takes fewer materials—per unit of electricity—to build a coal plant than a solar one, but then it takes *a lot* of coal to

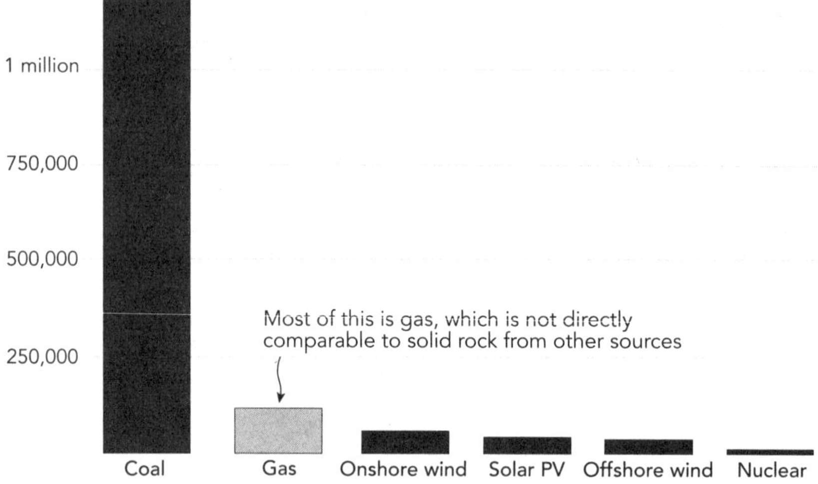

Figure 34.1
How many tonnes of material need to be mined for each TWh of electricity. This data includes construction materials and fuel, and the rock moved in the process of mining. The quantity shown for gas is the amount of gas extracted, which is not directly comparable to rock for coal, renewables, or nuclear.
Data source: Wang et al., "Updated Mining Footprints and Raw Material Needs for Clean Energy," 2024.

continuously run it. Once a solar panel or wind turbine is built, no more materials are needed. The fuel is free.

The bottom line is that there is not going to be a massive expansion of mining in a low-carbon world. There will be less.

WHAT WE NEED TO DO

The fact that mining quantities will go down in a low-carbon world shouldn't stop us from trying to improve things even more. Here's what we can do to make this transition in a more material-efficient and responsible way.

- **Build responsible supply chains.** While the quantity of mining might go down, there are still real concerns when it comes to other environmental and social impacts (see Questions 35, 36, and 37). We need to ensure that we're building fair and safe supply chains. Solving climate change while exploiting meager wages and child labor is not acceptable. This means working with governments and mineral suppliers to make sure that standards are enforced and workers are paid a decent wage.

- **Invest heavily in mineral recycling.** Better to use the stuff we've already pulled out of the ground than to dig up more. If we can ramp up recycling in the next few decades, we will need far fewer minerals than all the studies above think we will.
- **Make our solar panels, turbines, and batteries more material-efficient.** Combine recycling with improvements in material efficiency and we could slash our future demand for minerals. If manufacturing becomes more material-efficient, the minerals from the solar panel in your roof might be repurposed into two solar panels. That's because the technologies of the future should need much less lithium, silicon, copper, cobalt, and other materials than they do today. This is already happening. As we saw in the previous question, the amount of silver in a 2018 solar panel could be used to manufacture two solar panels now. The silver from a 2010 one could make *five* panels today. A future is possible where the amount of new materials that need to be mined each year is relatively low. This won't be for many decades but could be feasible in the medium to long term.
- **Reduce demand by investing in public transport and livable cities.** Car sharing and investments in public transport would make a big difference because electric cars do need more mining than gas ones. Combine these measures, and the total amount of minerals we need to extract for cars could more than halve.

THINGS TO BEAR IN MIND

Here are four things to be aware of.

1. The numbers included in this question's chart include all the ores and rocks that need to be moved or extracted to get those minerals out. Minerals such as copper, lithium, and iron are often found in rocks at very low concentrations, which means you need to mine a lot more

rock than you might expect from the materials alone. For example, the concentration of copper in ores is around 0.6%. To get one kilogram of copper, we need to mine 160 kilograms of ore, and 510 kilograms of rock need to be moved. This is also true for fossil fuels. This conversion makes a big difference to our numbers.

2. There are many charts floating around the internet that can easily fool you into thinking renewables need far more materials than fossil fuels. These charts are often out of date: The most popular one was published in 2015 when the market for solar and wind was almost nonexistent. These technologies have become much more efficient over time. These charts will also often ignore the coal and gas burned in the power plant itself. They'll show the materials used to build the plant but miss the fuel, which dwarfs the materials needed for renewables and nuclear. Don't be taken in by these charts.
3. The comparisons for natural gas are a bit more complicated than coal. To produce one terawatt-hour of electricity, we need around 125,000 tonnes of gas. The difference is that, when we're comparing the minerals and materials for coal, renewables, and nuclear, we're mostly comparing the amount of *rock* that has to be mined. Natural gas is, of course, a gas rather than a solid, which makes direct comparison more difficult. There is some disruption for gas: the rock extracted during drilling, the materials used for pipelines, well casing, and drill pads. But I haven't seen good and comparable estimates on the total amount of rock that's moved or used for this process.
4. Leading energy analysts estimate that in 2040—around the peak of our scale-up of low-carbon energy—we'll be using around 25 to 50 *million* tonnes of minerals to build solar panels, turbines, the transmission network, and electric vehicles.[2] Mining for clean electricity production will fall as the world moves from coal to solar and wind.[3] When it comes to electric vehicles, mining might increase slightly compared to gas cars. But when we combine electricity and transport, the amount of ore that would be mined would still fall by a lot.[4]

So, we might need to mine tens to hundreds of millions of tonnes every year at the peak of our clean energy rollout. Millions? That sounds huge!

Until we compare it to the 15 *billion* tonnes of coal, oil, and gas that we extract every single year. Without an energy transition, we'll extract at least 15 billion tonnes of fuel year after year: Once you've burned it, it's gone and can't be recovered. It's nonstop extraction. This is not the same for low-carbon technologies. We need an initial scale-up period to build out lots of infrastructure. But we will eventually reach a much steadier state where materials can be recycled or refurbished into new technologies.

QUESTION 35

What about human exploitation in mineral supply chains?

Answer: Minerals for many industries (including fossil fuels) are mined under unsafe and exploitative conditions, but clean energy can be an industry that changes things.

Cobalt is often called the "blood diamond of batteries." Three things are absolutely true: Electric vehicles are now the biggest users of cobalt, three-quarters of the world's cobalt is produced in the Democratic Republic of Congo (DRC), and mining in the DRC is often a brutally exploitative industry.[1] Hundreds of thousands of people work in the informal mining sector. Some of them are kids, and the pay is extremely poor: A worker will earn a few dollars a day.[2] Injuries and deaths from collapsing mines or tumbling rocks are common. Chemical pollutants lead to respiratory problems, and workers' eyes are often red or discolored through exposure to toxic materials.

This is unacceptable, and we need to do something about it. But it's worth pointing out that electric vehicles aren't the only driver of this exploitation. If you've got a mobile phone in your pocket, are reading this on Kindle, or want to send me an angry email on your laptop, then you're probably also using a battery that has cobalt from the DRC in it. Parts of the fossil fuels supply chains use cobalt too, including as catalysts in combustion engine cars. This is not a new problem. It's just that most people didn't know about it or didn't care. Now that electric cars have entered the scene, people seem to have found their conscience. Figure 35.1 lays all this out.

Now, what do we do about the cobalt problem? We could either stop using cobalt in batteries or improve working conditions in supply chains.

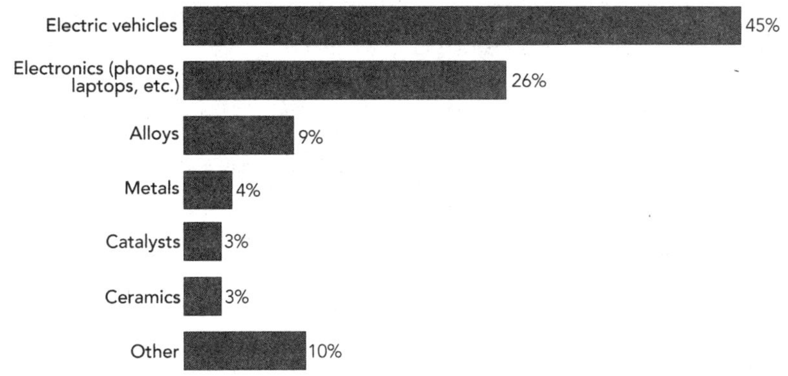

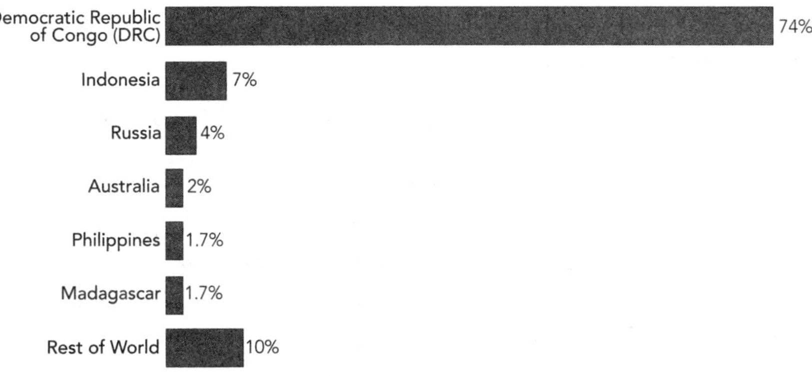

Figure 35.1

Cobalt use in 2023, and where it was mined. Electric vehicles use the most cobalt, but phones, laptops, and other electronics have been using cobalt for decades.

Data source: US Geological Survey and Cobalt Institute, 2024.

The first option looks far more likely as cobalt-free battery chemistries are becoming more popular. More than half of Tesla's new cars now contain lithium iron phosphate batteries, which have no cobalt (but still contain lithium, another problem mineral). Leading Chinese brands—such as BYD—are investing in sodium-ion batteries, which have no cobalt or lithium. Each year, Bloomberg New Energy Finance publishes forecasts about how much minerals we'll need in the electric car market in the future. Every year its cobalt projections tumble.[3] In its 2019 publication, it estimated we'd need more than 300,000 tonnes in 2030. Just three years later, it had cut this forecast to less than 150,000 tonnes. There is a possible future where cobalt isn't used in electric car batteries at all.

It's not just cobalt that we're worried about, though. As we move to low-carbon energy, we'll need a range of different minerals: lithium, manganese, copper, nickel, and rare earth metals (a group of 17 elements that aren't particularly rare but are critical elements in many electronic devices). These minerals are often concentrated in a small number of countries. Many are low- or middle-income countries with high levels of corruption and poor levels of governance. Almost half of the world's mineral reserves are in countries with "weak" or "poor" governance scores.[4] Ninety-four percent of rare earths, 70% of cobalt, 59% of nickel, 41% of copper, and 34% of lithium are located in these countries.[5]

In short: Many of the minerals we need for low-carbon energy are in countries where people are vulnerable to exploitation. And we can't put all the blame on the mining country being corrupt or poorly governed. Richer countries importing the stuff are also complicit: They know what's happening and aren't doing anything to change things.

Like cobalt, there might be some high-risk minerals that can be replaced with adjustments in technology. But we can't assume this for every mineral, so we need to make sure these minerals are more responsibly sourced too.

WHAT WE NEED TO DO

We have a few options to reduce the social exploitation within supply chains for critical minerals.

- **Switch to alternative clean energy technologies.** Batteries can come with different chemistries: a mix of different elements that are used to store and release energy. Over the last few decades, lithium-ion batteries—with a chemistry that contains cobalt—have become the most popular in electronic devices and electric cars. But as already mentioned, some leading car manufacturers are now using cobalt-free batteries instead. Tesla has switched many of its cars to lithium iron phosphate batteries that have no cobalt. Several Chinese companies are working on sodium-ion chemistries that have no cobalt or lithium. As a consumer, buying an electric car from a cobalt-free manufacturer means you're supporting companies moving away from highly exploitative minerals.
- **Invest in mineral recycling.** Sorry to sound like a broken record, but this is important. Exploitation happens when you need more and more extraction. The recycling of batteries, solar panels, and wind turbines will be essential going forward (and it's already happening). But it won't solve the problem in the near term: We're in the early stages of the transition, and we don't have enough recycled material yet. We'll need other solutions during this initial ramp-up period.
- **Track where minerals are coming from in supply chains.** Most companies that use critical minerals in their products don't know exactly where they're coming from and can't verify whether they have been sourced ethically. This needs to change. If consumers care enough, sourcing minerals from safe and fair mines can make a company stand out against its competitors too. Some governments are starting to put regulations in place to force companies to take responsible sourcing more seriously. In 2023, the European Union implemented its "Critical Raw Materials Act," which forces suppliers to regularly audit their supply chains. It also sets limits on how much of each mineral can be sourced from a single country.

THINGS TO BEAR IN MIND

Boycotting minerals from countries or supply chains that don't treat workers fairly seems like an obvious solution to the problem. If child labor is used in the DRC, then refuse to buy cobalt from the DRC. Or find a technological fix so you don't need cobalt at all. But it's a bit more complex than that.

The DRC is one of the poorest countries in the world. More than 60% live on less than $2 per day; 80% do not have access to electricity or safe drinking water. Mining accounts for 90% of the country's exports and is a vital source of income for many people in the country.[6] They might be teetering on the poverty line, but take that job away and they plunge below it. A move to cobalt-free batteries is not going to help them.

The DRC and other poor countries have huge stocks of resources that richer countries want and need. In an ideal world, they would reap the economic rewards. It would be great if these resources lifted huge numbers out of poverty, gave people access to well-paid jobs, and critical minerals became an engine for development rather than an exploitative burden.

Improving and reforming mineral supply chains would deliver greater benefits to the world's poorest than bypassing these minerals completely by switching to cobalt-free batteries. If miners in the DRC can work safely and get a decent wage, then they can take their kids out of the mines and pay for them to go to school. "Just improve conditions in supply chains" might sound like a naive solution, but it's not impossible. The world has raised standards in industries such as textiles and construction (although it's far from being a solved problem). Public pressure played a key role in these reforms, which is why—as citizens and consumers—our voices (and wallets) are important to achieve the same for mining.

QUESTION 36

Won't we become dependent on a few countries, just like we did with fossil fuels?

Answer: Unequal distribution of critical minerals puts the energy *transition* at risk, but not energy security; every country has free sunshine and wind.

A very small number of countries dominate the world's mineral supply chains. You can see this in figure 36.1. As we saw in the last question, more than 70% of cobalt is mined in the Democratic Republic of Congo. More than 60% of rare earths and graphite are produced in China. More than 70% of platinum comes from South Africa. The *refining* of these rocks into usable materials is even more concentrated. China completely dominates, refining more than 80% of rare earths, and dominating the cobalt, lithium, and copper chains too.[1]

People therefore argue that moving to clean energy puts us at a much higher risk of energy insecurity; just one conflict could cut off most of the world's energy supplies.

Of course, volatile energy markets are nothing new. They have been the standard for a world run on fossil fuels. When Russia invaded Ukraine in 2022, millions were pushed into energy poverty.[2] Up to 50,000 people across Europe alone died because of cold conditions and a lack of heating.[3] High energy prices meant high food prices, as the cost of fertilizers shot up. This is just one of many examples where our dependence on fossil fuels makes many countries energy insecure.

The risks of low-carbon energy and fossil fuels are not the same. Coal, oil, and gas are *fuels*; you need a continuous supply of them to keep the lights on and boilers running. If supplies are cut off, you have no power. This is a massive risk because most countries are importers of fossil fuels (see figure 36.2).

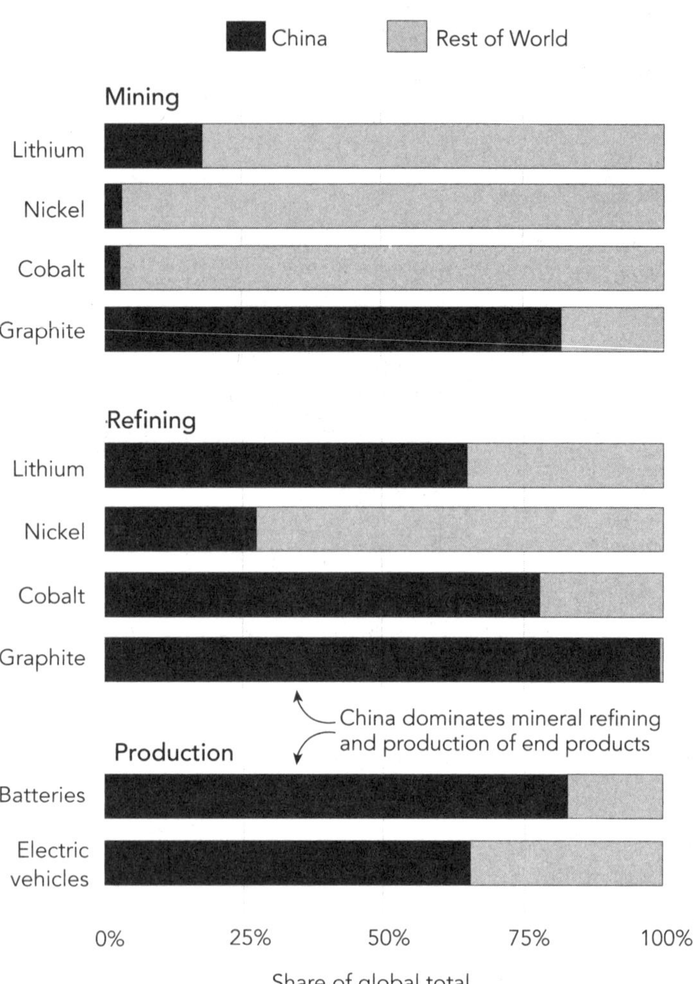

Figure 36.1

Who supplies the world's low-carbon technologies? Relying on a few countries for minerals increases the geopolitical risk of the energy transition, but not necessarily the risk of energy supplies once those technologies are installed.

Data source: International Energy Agency, 2024.

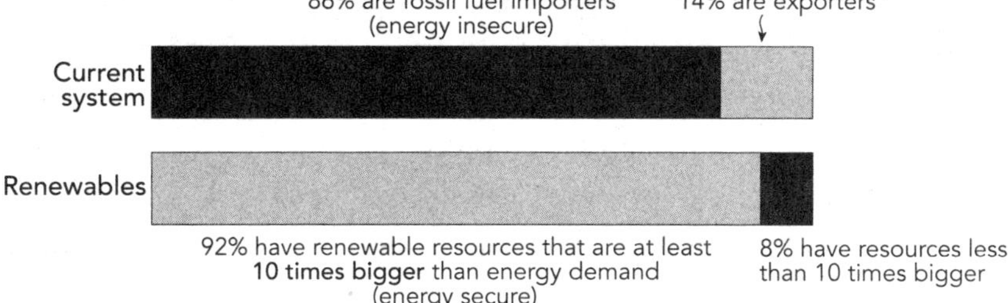

Figure 36.2

Share of the world population that are importers or exporters of fossil fuels, compared to a renewable system, where most of the world has control over its fuel supply. Most countries are fossil fuel importers; in a world run on renewables, they have their own resources.
Data source: Carbon Tracker and Kingsmill Bond at RMI.

For solar and wind, the fuel is free. No one can block out the sun or stop the wind blowing; other countries can't cut off your energy supply. They *could* cut off your supply of construction minerals, but those are used to build *new* plants and batteries; you don't need them to keep your existing ones running.[4]

In a fossil fuel economy, a conflict that throttles your supplies leaves you without energy. In a low-carbon one, your existing power plants and cars can keep running as normal. You might be temporarily blocked from building *new* infrastructure, but your existing plants can keep going. Fossil fuel dependency is a risk for energy security. Mineral dependency is a risk for the energy transition. It might halt or slow down the transition but won't stop it completely. Countries would need time to open mines and build new mineral refining plants at home; but this could be done in a matter of years.

WHAT WE NEED TO DO

High prices, unstable markets, and geopolitical tensions might not cut off energy supplies, but they could slow down the energy transition, and we don't have time for that. If we want to make the energy transition fast and smooth, we need to diversify and reduce the risks. Here's what we can do.

- **Countries can expand their search for minerals at home.** The more countries that produce minerals, the better. This reduces tension between different regions. Many countries are investing in mineral exploration to find their own supplies. The US seems to find another huge lithium deposit every week. A big part of its Inflation Reduction Act package focused on bringing mineral processing and manufacturing back "home." Australia keeps finding new deposits of rare earth metals. Norway recently found a giant phosphate deposit and Europe's largest source of rare earth metals. More of this, please.

- **Use new technologies to help us find more mineral deposits.** Artificial intelligence and improved imaging techniques can help us unearth deposits that we would have previously missed. There are a bunch of start-ups—such as KoBold Metals and HyperSpectral—working on cool AI innovations that can use imaging and data patterns to predict where new mineral deposits might be hiding. Of course, AI is also energy intensive, so we want to make sure we're using it responsibly and for the high-impact innovations that could make the energy transition easier, not harder.
- **Take advantage of technological innovations to avoid high-risk minerals.** There is a range of possible technologies that can be used for solar panels, wind turbines, or batteries. As we have already seen, alternatives are being developed to avoid using cobalt and lithium in batteries, so if there are supply issues, manufacturers have something to switch to. We see the same substitution effect in electricity grids: When copper prices are high, aluminum can be used instead. This flexibility is less available in fossil fuel economies, so it gives us the chance to make low-carbon energy systems more adaptable to geopolitical risks.
- **Invest in mineral recycling and repurposing.** Yes, this again! I hope the world can get as good at recycling minerals as I am at recycling recommendations. Fossil fuels are burned and can't be recovered. You need to dig out more coal again and again. For low-carbon technologies, there is an initial ramp-up period to build the infrastructure, but when the panels or batteries reach the end of their lifetime, the materials can be recovered and recycled. Countries can start to build a more circular model of material use, repurposing what they already have rather than continually buying more from others.

QUESTION 37

Doesn't mining minerals for clean energy use too much water to be sustainable?

Answer: Clean energy tends to use less water than fossil fuels, but we need to make sure the transition doesn't exacerbate local water stress.

It would be ironic if our solutions to climate change made water stress worse, not better. This is a genuine concern. Producing minerals for solar panels, wind turbines, batteries, and electric cars uses water. Often a lot, and in places where water supplies are under pressure. You might have seen pictures of the large evaporation pools in South America, where miners have been extracting lithium for batteries in areas of water stress.

When comparing the water use of electric and gas cars we need to include the water used to mine the minerals and manufacture the car itself *and* the water used to produce electricity or gas to run it.

It is true that an electric car uses more water in many countries today. You can see this in the chart. An electric car in China consumes 262 cubic meters (m^3) of water across its lifetime, compared to 137 m^3 for a gas one.[1] Almost twice as much. In the United States, an electric car consumes around one-third more water.

As you can also see—by comparing the first two bars in the chart—electric cars use more water than gas cars in China. This is because the electricity powering electric cars is still coming from fossil fuels. But if the electricity was coming from solar or wind, they would consume about two-thirds *less* water than a gas car.[2] Moving to solar and wind makes such a big difference because most of the water used by electric cars—around 82% in the US and China—comes from generating the electricity used to power

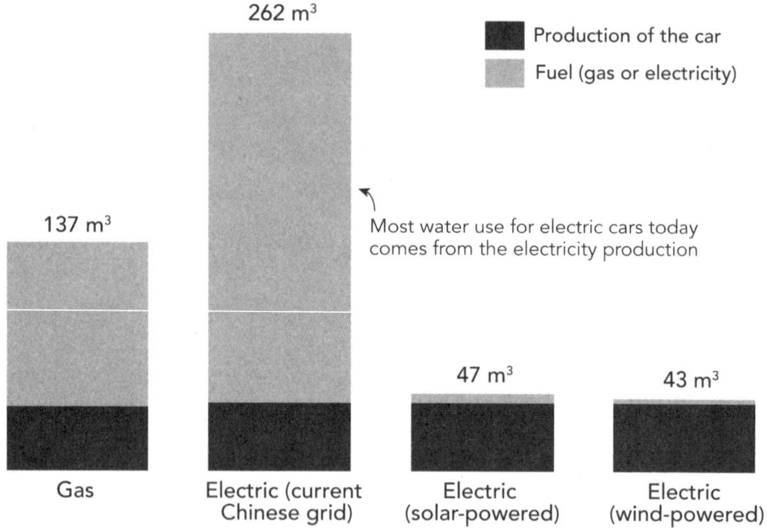

Figure 37.1

How much water cars use in their lifetime. Electric cars use more water if the electricity grid has lots of fossil fuels. When powered by solar and wind, water use is much less than gas.
Data source: Wang et al., "Life Cycle Water Use of Petrol and Electric Light-Duty Vehicles in China," 2020.

them, not to build the body of the car or the battery. And solar and wind use *a lot* less water than fossil fuels or nuclear power: around 5 to 10 times less than natural gas, and 10 to 100 times less than coal *per unit* of electricity.[3] Coal, gas, and nuclear plants need high volumes of water to keep the plant equipment cool. The only way our energy system would consume *more* water in a low-carbon world is if we heavily depended on nuclear and bioenergy, which are the most water-hungry.

Water use in the United States is already falling as it moves from coal to natural gas and then to solar and wind, and water use globally will go down too as we move to low-carbon energy.[4]

WHAT WE NEED TO DO

The fact that *global* water use will decrease as we move to sustainable energy doesn't mean that *local* water stress from mining the minerals for these technologies isn't a real problem (see "Things to bear in mind"). But there are things we can do to reduce the pressure on local supplies.

- **More public transport and car sharing.** This is obvious. Fewer cars mean fewer resources. As individuals, we can help with this by opting not to buy an electric car if we don't need one.
- **Reduce mineral demand through efficiency and recycling.** Even if we can reduce car ownership, the world is still going to need a lot of these minerals. We can cut mining demand massively by reducing the amount of lithium, nickel, or copper needed *per* solar panel or battery; and recycling old materials so we don't need to dig up more.
- **Diversify mineral supply chains so production isn't hyper-focused on water-stressed areas.** Diversifying our supply chains is not just good for water use but makes sense for geopolitical reasons too. Many countries are already investing in mineral exploration at home: You'll find several examples of this in Question 17.
- **Focus on mining technologies that conserve water.** There are ways to extract the same amount of minerals with less water. This should be a priority for mining companies, but also processing and technology manufacturers along the supply chain. They can invest in water repurposing and reuse. Water can often be extracted from "tailings"—the waste rock and residue that mining leaves behind. Some are experimenting with new ways of extracting minerals. "Direct lithium extraction" captures lithium directly from the brine, without having to evaporate all the water. Microbial technologies might also be able to reduce levels of toxins and pollutants in wastewater, so that it can be used within the mining site. Such innovations will be crucial to reduce pressures and make sure mining is done more sustainably.

THINGS TO BEAR IN MIND

Local water stress from mining is still a real problem. Around 16% of the world's mines or deposits of critical minerals are in areas of high water stress.[5] It's an even bigger problem for copper and lithium: Around half of the world's supply is produced in areas with extreme or high water stress.

Take the example of lithium. In some mining sites—particularly in Australia—lithium is produced by digging hard rock out of the ground. But lithium can also be produced from brine: Salty liquid from underground is pumped to the surface and left in open pools to evaporate, leaving lithium and other minerals behind when the water has disappeared. This method is most common in the "lithium triangle" of Chile, Argentina, and Bolivia, where water stress is already very high.

It's not the *total quantity* of water used for mining that's the problem. Mining uses a tiny fraction of the world's water, which is mostly used for farming. Even in water-stressed Chile, mining uses just 4% of water; 72% is used by agriculture.[6] Within the energy sector, extracting oil and producing electricity uses far more water than the mineral mines themselves.

The problem is that *all* the world's critical minerals are being produced in a handful of locations. In some northern regions of Chile, more than half of fresh water is used for mining. That creates highly localized areas of water stress, and intense conflict for local communities who rely on freshwater resources to live. If more countries—and more sites—were producing these minerals, we could alleviate pressure on some of the current hotspots.

One final thing to consider is that mineral mining and processing don't only *consume* water: They also risk contaminating surrounding water supplies. Indigenous communities in Chile and Argentina have reported the contamination of fresh water from nearby lithium mines. These activities need to be tracked properly so that mining stops long before there is any risk of groundwater contamination for local communities. This is relatively easy with modern monitoring technologies and must become the norm within mineral supply chains.

VII

HEATING (AND COOLING)

More than half of the world's energy used in buildings is for heating: It accounts for more than 10% of global carbon emissions.[1] But it's an area where we already have good solutions.

The front runner is replacing gas or oil boilers with heat pumps. Heat pumps are kind of magical: They *produce* more energy than they *use*. In an electric heat pump, every unit of electricity typically produces three or four units of heat because they harvest heat from the air, ground, or water around them.

Compare that to an electric space heater—like a radiator. The maximum efficiency you can get is 100%: one unit in, one unit out. That's because they're just turning electricity into heat. Heat pumps are also three to five times as efficient as gas boilers. It's an incredibly neat piece of engineering.

But heat pumps are often unpopular: People claim that they don't work in the cold, are useless without insulation, and are too expensive. These perceptions are blocking our best shot at moving to low-carbon heating.[2]

Let's look at these claims to see if there's any truth to them.

QUESTION 38

Aren't heat pumps hopeless in the cold?

Answer: Heat pumps are less efficient in cold temperature but *do* work; in fact they're most popular in some of the coldest climates.

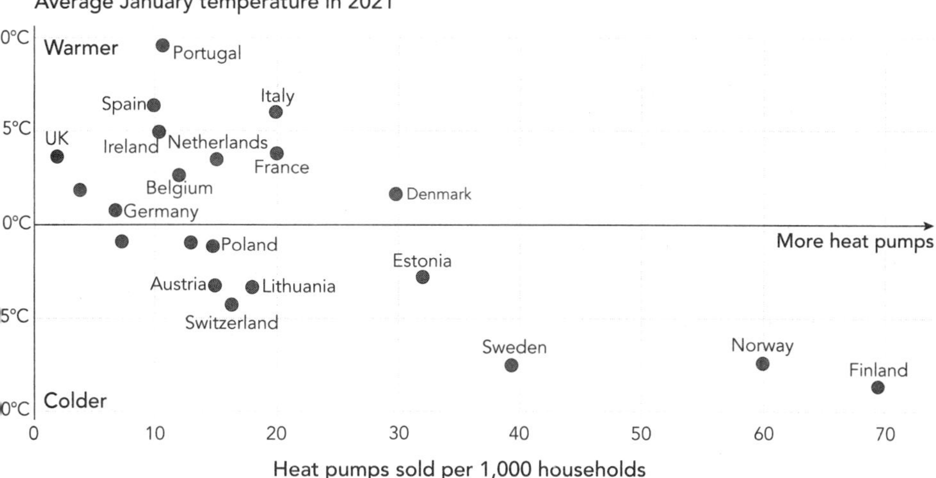

Figure 38.1
Heat pump popularity in Europe. Sales of heat pumps per 1,000 households are measured against average January temperatures.
Data source: Heat pump data from the European Heat Pump Association.

Heat pumps are most popular in the coldest countries in Europe, so it would be ironic—and a bit useless—if they didn't work in cold climates. Thankfully that's not the case.

We don't even need to look at the technical performance of heat pumps to figure this out. We just need to look at *where* heat pumps have already been installed. You can see in figure 38.1 that snowy Norway, Finland, and Sweden have the most heat pumps per million households in Europe.[1] You'll also find many heat pumps in Alaska, where temperatures can get below 20°C in winter.

The effectiveness of heat pumps is measured using a term called the "coefficient of performance" (COP), which counts how many units of heat are produced for every unit of electricity we use. So, if our COP was 2, then we'd produce two units of heat energy for every one unit of electricity used. A COP of 3 would mean we get 3 units out.

The efficiency of heat pumps *does* fall when it's colder. This isn't surprising. But most heat pumps still perform well. Between 5°C and –10°C, the average COP was around 2.7. That's pretty good. One study looked at the performance of 550 heat pump systems in Switzerland, Germany, the United Kingdom, Canada, the United States, and Germany.[2] These COP values ranged from 3.3 in Canada, to 2.5 in the UK, down to 2.4 in China. Heat pumps are more efficient than gas boilers (which are only around 70–80% efficient: For every unit of energy used we get 0.7 or 0.8 units back) or electric radiators (which can get close to 100%: one unit in, one unit out).

Enough of the technical studies: What do people who have heat pumps say? The innovation agency Nesta interviewed more than 2,500 heat pump users in Britain to see if they were happy with their systems.[3] Three-quarters said they were happier—or just as happy—compared to their previous system. And people were just as satisfied as those with a gas boiler. You'll often hear the claim that heat pumps don't work in older, leaky houses. But that's not true. People were happy regardless of whether they lived in Victorian or modern housing.

Whether we look at technical performance data or user satisfaction, it seems clear that heat pumps do work.

WHAT WE NEED TO DO

There are a few things we need to focus on to make sure that electric heat pumps can work to their full potential *and* people can stay affordably warm throughout the winter.

- **Insulate homes to cut down on energy bills.** Many people argue that heat pumps are useless if homes aren't insulated. Of course, having insulation helps a lot. But this is true for *any* heating system: heat pump, gas boiler, or otherwise. Yes, our homes should be insulated to save energy and make heating more affordable. Heat pumps alone are not going to solve energy poverty. But most homes don't need to wait for insulation *before* installing a heat pump.
- **Invest in high-quality training programs for heat pump engineers.** Having a skilled workforce to fit heat pumps is crucial. If they aren't installed properly, they *won't* be effective. In the study above, some heat pumps achieved far less than the average efficiency, and poor installation was probably to blame. Heat pumps are more complex to install than a gas boiler; they require engineering, plumbing, pipe-fitting, and refrigeration skills rolled into one. If the world is going to decarbonize its heating systems over the next few decades, we're going to need *a lot* of skilled heat pump installers. It's a complex job, and one that deserves to be well paid.

In the next five years, my own country, the UK will have to increase the number of engineers five- to tenfold.[4] By 2030 it could need as many as 30,000. Today it has just 3,000. The government is investing in upskilling programs, but it's still too slow.

QUESTION 39

Aren't heat pumps much more expensive than a gas boiler?

Answer: The running costs of heat pumps are often lower than gas boilers, but upfront costs really *are* too much.

The critics are right: Heat pumps are expensive. They are usually cheaper to *run* than gas boilers, but the upfront costs are steep. Without government support, they'll be out of reach for most households. Even with generous grants, many will struggle. Let's look at both costs individually.

First, the price of buying and installing a heat pump. Upfront costs can vary a lot depending on the efficiency and quality of the heat pump you're installing, how much home retrofitting is needed, and labor costs. But in the US, installing an efficient heat pump is typically around $10,000 more expensive than a gas furnace. You can get a low-efficiency one for cheaper, but it will cost more to run and will reduce some of the environmental benefits of making the switch. Previously, the Inflation Reduction Act offered financial support, with a $2,000 tax credit, and an additional $11,500 in rebates for low-income households. These did significantly improve the economics of heat pumps in the US, especially for those on lower incomes. As of July 2025, Trump's new "Big Beautiful Bill" slashed these credits. Some states might still choose to provide incentives at a local level, but these national cuts will undoubtedly slow the transition to low-carbon heating and reduce the odds that capital costs of heat pumps improve.

The upfront economics are similarly difficult in the UK: The average cost of getting an air-source heat pump is around $16,000, compared to $4,000 for a gas boiler.[1] People are not going to pay $12,000 more to install

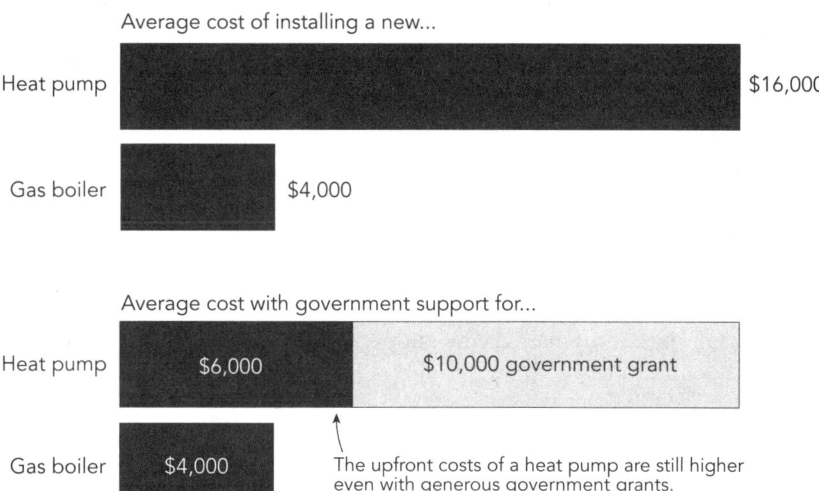

Figure 39.1
Cost of installing a heat pump or gas boiler in the UK.
Data source: UK government, MSC and Jan Rosenow in Carbon Brief. Converted to US dollars using a conversion factor of 1.3.

a heat pump, so the government offers households a grant of up to $10,000. Many other countries across Europe offer similar subsidies. That still leaves a $2,000 shortfall for many households. Some will be able to afford this, especially if they can repay it with lower running costs, but many lower-income families won't.

You can see this inequality in real buying behaviors. Nesta ran a survey across 2,500 heat pump and 1,000 gas boiler users in Britain. Heat pumps are rare in the most deprived areas, with adoption rates rising toward the richest.

The economics of *running* a heat pump is a bit better. One crucial factor is the difference in the price of electricity and gas. In the US, this can vary a lot from state to state. Heat pumps are more popular in southern states like Texas and Florida, where electricity is cheap and natural gas is more expensive than the national average. Since these homes are also often set up for electrification with air conditioning, it makes the installation process much easier. While the *amount* that households will save varies a lot by state, most will find that they have lower energy bills with a heat pump than a gas furnace (and much lower than with oil heating). This is also true elsewhere: The average household in the US, Europe, South Korea, and Japan reduced their heating costs by switching to a heat pump in 2021 and 2022.[2]

Unfortunately, in the UK the cost benefits are smaller because electricity is much more expensive than gas: currently around four times as much.[3] You might think that this means running a heat pump would be four times as expensive too. But remember that an electric heat pump is much more efficient. If a heat pump has an efficiency—measured as the number of units of heat you get out per unit of electricity in—of three or more, then it will cost the same to run as a gas boiler.

Some companies in the UK offer much cheaper electricity tariffs for heat pumps, making them cheaper to run than gas boilers, even when their efficiency is less than 2.5. Combine an efficient heat pump with one of these tariffs, and gas boiler users could halve their heating bills if they made the switch.

All in all, the lifetime cost of an electric heat pump—including the cost of buying and running it—will be lower than a gas boiler in most countries *with subsidies* that help with the upfront costs. Consumers might need to invest more to cover the installation, but heat pumps will save a few hundred dollars in running costs for most households each year. Over the lifetime of a heating system, that will more than cover any additional upfront costs. But what about costs *without* financial support? A large study by the National Renewable Energy Lab in the US found that heat pumps would be the cost-effective heating choice in 60% of the country—around 65 million households—without any tax credits or subsidies.[4] Again, that's over the lifetime of the heating system. Heat pumps were cheaper to run in 86% of homes if they installed a medium-efficient pump, and 95% with the highest-efficiency ones. But these savings were not always sufficient to offset the higher upfront cost.

So, total cost for governments could be steep as they're footing the bill for these extra installation costs. There's no point in denying that decarbonizing heating could come with an initial hefty price tag. But it's worth remembering that electric heat pumps are the *cheapest* way to decarbonize the world's heating: cheaper than electric heaters and hydrogen.[5]

WHAT WE NEED TO DO

If the economics of heat pumps stay as they are now, this transition is going to go incredibly slowly. Here are things we can do to speed things up and make it a more equitable one.

- **Improve government grants, especially for lower-income households.** Many governments already offer grants, but more generous ones for lower-income households would make a big difference. In many countries, even with government grants and incentives, the average heat pump still costs several thousands more than a gas boiler. This also means that these households are priced out of the heat pump market and can't benefit from the energy savings of *running* one.
- **Make electricity more affordable–gas less so.** If you want people to electrify their heating systems, you need to make electricity cheap. Instead, in some countries most energy taxes are placed on electricity, rather than gas, making electrification *less* affordable. Moving these payments to general taxation, or taxing gas instead, would make electric heat pumps more appealing.
- **Offer more attractive tariff schemes.** UK companies like Octopus and OVO are making a big difference with their better electricity tariffs for heating. This dramatically lowers the running cost. More of this, please.
- **Innovate in heat pump technologies to drive down the cost.** Prices *are* falling, just not fast enough to be competitive with gas boilers. Upfront costs are expected to fall by 25–30% by 2030.[6] That could cut the cost by $4,000 to $5,000. This could make even the upfront costs much more competitive for some households; but for many, a heat pump will still be more expensive to buy than a gas furnace or boiler. We need to make this happen faster, and provide support for the lowest-income households.
- **Use heat pumps to provide both heating and cooling.** The US is in quite a unique position, because many households need both heating and air conditioning (in the UK we have heating, but very few homes have A/C). If a home already has air conditioning, they could replace it *and* their gas furnace with a heat pump that does both. So, there's just one dual-system. That tends to improve the economics.

QUESTION 40

How can the world deal with the increasing demands for air conditioning?

Answer: More people will get air conditioning and that's a good thing, and with efficient A/C units we can keep energy use in check.

In a world that's getting hotter, the ability to stay cool moves from a luxury to a basic human right.[1] Air conditioning (A/C) is a lifesaver. Not only that, it helps children learn properly at school, allows people to sleep in comfortable conditions, and lets them work to their full potential.

There are around 2 billion A/C units in the world today.[2] By 2050, this number could triple to 6 billion. This is because cost is currently what's stopping many people from getting air conditioning, and incomes will rise in the next few decades, especially in low- and middle-income countries.[3] This is undeniably a good thing. Living standards will improve *and* more people will be protected from life-threatening heat. But A/C is already responsible for 3–4% of the world's greenhouse emissions, and guzzles up around 8% of the world's electricity.

Some people, then, view A/C as a luxury that the world can't afford. The energy and climate impacts are too high, and we should convince people not to use it. Electricity demand for cooling could triple by 2050, but by making sure that people have more efficient A/C units we could halve electricity demand—a saving equal to the electricity consumption of the European Union.

There are lots of things we can do to reduce energy demand for A/C without sacrificing people's right to stay cool (see "What we need to do"). But the biggest barrier is cost. Many people can't afford an air conditioner

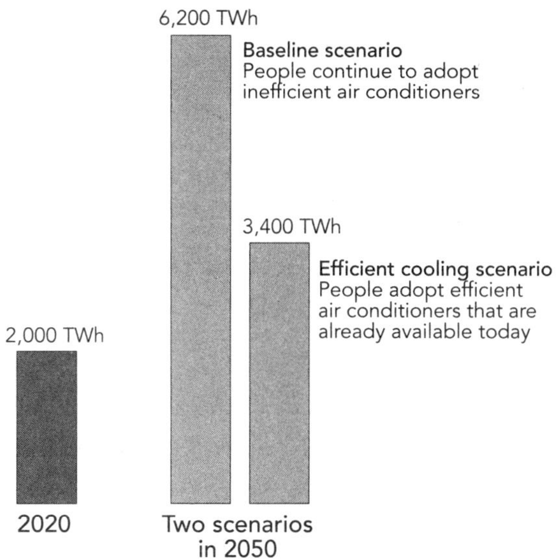

Figure 40.1

The future of electricity demand for air conditioning: the amount of electricity "saved" in 2050 if efficient air conditioners are used is equal to the electricity use of the European Union.
Data source: Based on data from the International Energy Agency and Future of Cooling.

at all. For others, when they *can* afford their first A/C unit, they opt for inefficient models that are cheaper. I looked at a range of air conditioners from one of India's biggest brands. The cheapest three-star efficiency option costs around $350. The cheapest five-star one was $100 more expensive—that's a 20–35% premium that many households just can't afford. The real kicker is that the cheap inefficient ones are far more expensive to run. A family that buys a unit that's 50% more efficient will halve their electricity bills and save $2,000 over the following years.[4]

Making sure that people have efficient A/C units will not only help them keep cool affordably but will also make a huge difference to energy demand and the climate.

Something I come across often, and think we need to avoid, is seeing air conditioning as a self-indulgent luxury when it's a lifesaver. Keep in mind that *heating* globally emits around four times as much as cooling. People don't view this as a luxury; they view it as a necessity. In the UK, we're rightly outraged when people can't afford to heat their homes. They're defined as being in "fuel poverty." The need for cooling is no different. If someone living in heat can't afford air conditioning, they're living in fuel poverty too.

WHAT WE NEED TO DO

There's a lot that we can do to keep people cool and comfortable while keeping their energy use and energy bills lower.

- **Develop more efficient A/C units that can handle humidity.** A/C units need to be able to deal with humidity, not just extreme heat. You might expect that all the energy used by an air conditioner is to control the temperature, but up to half of its energy is used to control humidity, by removing water vapor from the air.[5] This is important because our bodies struggle to deal with heat when the air is humid; we lose our ability to sweat and thermoregulate. Removing some of that excess water from the air makes conditions more comfortable. The problem is that A/C units are often not tested and developed for high-humidity conditions, so they are not as efficient as they could be.
- **Provide clear efficiency labels on A/C units.** The average A/C unit that's sold today is less than half the efficiency of what's available. In Ghana, 85% of installed A/C units would rate as one star (the lowest rating) on typical efficiency ratings.[6] Some households buy less efficient models because of poor labeling and communication. If units aren't given an efficiency score on the label, how are consumers supposed to know? This is where policies and standards play an important role. Governments need to bring in minimum efficiency standards to get the rubbish technologies off the shelves and put a consistent labeling system in place. This has already been done for other home appliances—such as refrigerators, televisions, and lighting—and efficiency standards have been effective in driving innovation and consumer choices.
- **Reduce taxes for the most efficient A/C units or provide support for the poorest and most vulnerable.** Governments might be able to help. Maybe taxes can be reduced for the most efficient air conditioners. Or

low-income households can receive subsidies that help them with the capital cost.

- **Maintain A/C units to keep them running efficiently.** Replacing filters, unblocking vents, and cleaning coils in the unit can reduce energy demand by around 25%.
- **Combine air conditioning with other cooling techniques.** This is particularly important in cities where the "urban heat effect" can make the hottest days 3°C warmer, and a stifling 12°C hotter during the night. Building design and materials are crucial: Choosing the right ones can reduce cooling demand in hot climates by 10–40%.[7] Cooler buildings and well-designed cities could also be a lifesaver for those who can't afford air conditioning. They're a benefit for everyone.
- **Provide public energy efficiency campaigns.** Just as people are often advised to "turn down the thermostat" when it comes to heating, the same is also true for cooling. Some governments may put guidelines on how low the thermostat should go, the number of hours to run A/C, and *when* to turn it on. India, for example, changed the default setting on new air conditioners from 20°C to 24°C, which can save 20% of energy per year. How to do this while ensuring that people are kept cool enough, and still have control over their home electricity demand, will be the socially challenging part.

THINGS TO BEAR IN MIND

One issue that countries will need to deal with is *when* A/C energy is needed, because of the pressure it puts on the grid at any given time. In the US today, cooling can make up almost 30% of peak electricity demand on the hottest days. With better A/C units, this would fall to just over 20% in 2050. The US has dealt with air conditioning for many years now—and has money to invest in grid improvements and more efficient A/C units—so it's probably going to manage this increased demand somehow.

Contrast that with India. Today, cooling can make up 10% of electricity demand at peak times. By 2050, using inefficient A/C units demand would rise to almost 45%. That's going to be a massive strain on the grid. If it uses efficient units, it would be just 20%. It's a similar pattern across most low- and middle-income countries: Inefficient A/C units will put serious pressure on the electricity grid. With efficient ones, they might be able to cope.

What's helpful is that the hottest times will be when the sun is shining. Solar power is going to be crucial to meet this A/C demand comfortably and to provide cooling energy without a colossal rise in carbon emissions. If countries can harness the power of renewables in this way, they could provide cheaper and more equitable air conditioning to those who need it the most.

VIII

FOOD

Mention climate change and people automatically think about fossil fuels and energy. Food often gets overlooked. But one-quarter to one-third of our emissions come from our food system.[1]

This *is* something we can tackle. Not only can we slash our carbon emissions, but we can also free up land for forests to regrow, let wildlife return to old habitats, and suck carbon *out* of the atmosphere. We can undo some of our environmental damage. We just need to be realistic about what things make a difference.

The biggest change we can make is to cut back on meat and dairy. Animal products tend to have a much higher carbon footprint than plant-based foods. That means a shift toward more plant-based diets (which doesn't mean everyone has to go vegan tomorrow) could considerably reduce the climate impact of our diets.

But would we have enough land to feed everyone on more plant-based diets? Are meat substitutes really better for the climate? Aren't they too expensive and unhealthy? Let's look at the data.

QUESTION 41

Surely we don't have enough land for everyone to go plant-based?

Answer: If the world went plant-based, we'd need less cropland because we'd free up land currently used to produce animal feed.

Would the world have enough farmland if everyone switched to a plant-based diet?

Around two-thirds of the world's grazing lands are not suitable for growing crops. At least, not very productively. So, if crops couldn't be grown on that land, maybe we wouldn't have enough? Or we'd need to cut down lots of forest to make space for it?

Seems logical, but the data shows us that it's a poor argument. This is because so much of our crops are fed to animals, not humans. Only around half of the world's cereals go directly to human food.[1] The rest goes to animal feed or biofuels. Three-quarters of soybeans are fed to livestock.[2] Our typical soy-based "vegan" products, like tofu, tempeh, and soy milk use just 7% of the world's soy. If we add in soybean vegetable oils, it's still only 20%.

There's an obvious caveat when we're looking at how much food goes to humans versus animal feed. Animals do, of course, convert that into tasty meat and nutritious protein that humans ultimately eat. But they do it very inefficiently. For every 100 calories we feed a cow, we only get around 3 calories back in return.[3] The other 97 calories are burned just keeping the cow alive. For pigs, we get 9 calories back. For chickens, 13. Most of the crops we feed animals are, effectively, "wasted." The returns on protein are a little better, but most of the protein is still lost in the process. Grazing and croplands for animal feed take up three-quarters of the world's agricultural

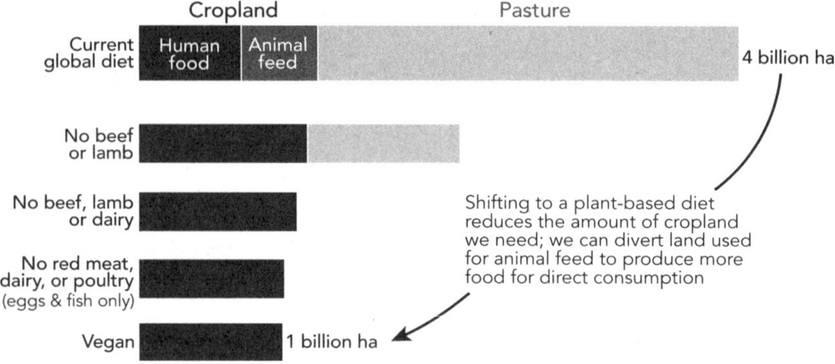

Figure 41.1

How much global land is needed for agriculture with different diets?
Data source: Joseph Poore and Thomas Nemecek, 2018.

land. Yet animal products supply just 18% of our calories and 37% of our protein. It's a crazily inefficient use of land.

A large study modeled various scenarios of global diets and found that a plant-based diet reduced agricultural land use by 75%.[4] You can see this in figure 41.1. That's not surprising, since we'd get rid of the huge amounts of land that animals graze on. But it also found that we'd need less *cropland* than we use today. We'd be trading land used for animal feed for land used to grow legumes, fruits, and vegetables that we can eat directly.

The data is clear: The world has enough land to support more plant-based diets. Now, I'm not deluded into thinking that the world is going to go fully vegan any time soon. Nor do I think that aggressively insisting on a completely plant-based world is the right approach to changing dietary habits. People will push back. But a reduction in global meat consumption could have significant impacts on reducing land use. So it's up to us to decide how far we slide the scale toward a meat-free world.

WHAT WE NEED TO DO

There are things we can do to support shifts from meaty to more plant-based diets, but there are also ways to cut agricultural land use even more.

- **Eat less meat and dairy.** Incredibly obvious, but this is something that *you* have a lot of control over. We get to choose what we eat three or more times a day. Again, this is not about going "cold turkey" overnight. If everyone ate a bit less meat it would have a much bigger impact than a few people going vegan.
- **Improve crop yields.** If we can increase yields then we can grow more food on less land. This means making sure farmers in every country have access to the best seeds, irrigation, fertilizers, and pesticides when needed. And it can come from scientists developing new crop strains: making wheat, rice, or corn varieties that are more heat- or drought-resistant.
- **Make proteins in the lab.** Rather than swapping beef and pork for peas and beans (which need to be grown outdoors), we could see a shift from the field to the lab or factories. Innovations such as lab-grown meat, precision fermentation—where we make proteins, fats, and vitamins from microorganisms—and other biotechnological advances could see us produce large amounts of food on almost no land.
- **Make fats and oils in the lab.** These could also be produced synthetically, reducing the need for palm oil plantations and soybean crops. Savor Foods is one start-up that's leading the way on producing synthetic fats, to give us tasty equivalents to butter and oils without using animals, land, or fertilizers.
- **Squash the notion that "natural is always better."** Public perception is one of the key things standing in the way of a transition from meat and fats to proteins and oils made in factories. People often think that "natural is good" and "synthetic is bad." There's no science to support this, but it's a feeling that many (including me, sometimes) struggle to shake off. We need to, though, if we're to achieve a widespread shift to alternative proteins. Figuring out how we change the entire culture of food is above my paygrade, but getting real and honest about the environmental and animal welfare impacts of our current food system is one place to start.

QUESTION 42

Aren't meat substitutes worse for the climate than meat because they use so much energy?

Answer: Meat substitutes do use energy for processing, but they still have a lower carbon footprint than meat.

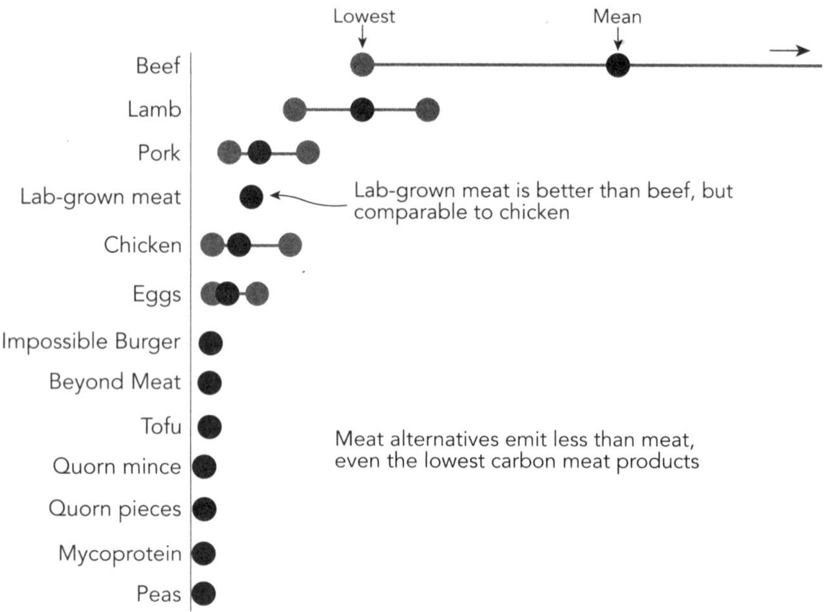

Figure 42.1

Carbon emissions of food products, ranked. Measured per 100 grams of protein and based on emissions across the whole supply chain. For animal products, the highest to lowest producers are shown from a large global study.

Data source: Poore and Nemecek, 2018; Sinke et al., 2023; industry reports.

Impossible or Beyond Meat burgers are processed, need lots of energy to be manufactured, and include a bunch of ingredients. Surely they can't be better for the climate than real meat from an animal?

In almost all cases, they are. Look at the chart, which compares the emissions of different protein products including all the emissions along the supply chain: from land use and farming to manufacturing and packaging. All of them have a *much* lower carbon footprint than beef and lamb, often more than 10 times lower. They're lower than pork. And also lower than most chicken products; only the most efficient chicken producers can achieve similar emissions. Meat substitute burgers and bacon strips *do* have a higher carbon footprint than classic plant-based foods like peas, beans, or lentils. But they beat meat almost every time.

Many people will say that comparing these products to the global *average* footprint for meat is unfair. They argue that *their* beef produced in Europe, or the US, has lower emissions. They're not wrong. But we can compare meat substitutes to the most efficient (i.e., the "best") farmers too. In figure 42.1, I've plotted not only the average but the lowest and highest producers. The result is the same, even if the gap is smaller. Meat substitutes nearly always win, even against the most climate-friendly beef or pork.

WHAT WE NEED TO DO

In addition to swapping meat for meat substitutes, here are a few things we can do to make our dinner even more climate friendly.

- **Move to clean energy.** What's exciting is that the meat substitutes we'll be eating in a decade will have an even lower carbon footprint than they do today. Not only because these products are in their infancy (the world has had cows and sheep for much longer than Impossible Burgers), but because our electricity grids will get a lot cleaner. When the emissions of our energy systems fall, the carbon footprint of processed meat substitutes will follow. Meat-free burgers and sausages are already much more climate friendly than meat. This will only get better.

- **Focus on the most sustainable producers of meat while meat consumption shrinks.** We need to get real: The world is not going to cut out meat tomorrow. To minimize emissions while we transition to more plant-based diets, we need to make sure that the meat the world *is* still eating comes from the most efficient producers. In figure 42.1, you can see huge differences in the carbon footprint for beef, lamb, and pork. Focusing on the most sustainable producers—and moving away from the most damaging ones—could cut emissions a lot.

THINGS TO BEAR IN MIND

One product that is still debated—because it hasn't been scaled up or hit the market yet—is lab-grown meat. Lab-grown meat has basically the same ingredients as meat we get from livestock today. Scientists take tissue and isolated cells from an animal or fertilized egg and grow the meat in a reactor instead. We get the same product, just without killing tens of billions of animals every year, and without the environmental impacts of raising them. Lab-grown meat does need quite a lot of energy. How "low-carbon" it is depends on how clean our electricity mix is. If it's powered by solar, wind, or nuclear then it's going to have a very low footprint. If it's coal powered then it'll be a lot higher. Several studies find that lab-grown meat will have a carbon footprint lower than most meats, and still lower than beef or lamb even if it's powered by fossil fuels.[1]

"Lab-grown meat could be 25 times worse for the climate than beef," declared a *New Scientist* headline, catching the attention of the media.[2] But this is unlikely. The study looked at the footprint of lab-grown meat under two different scenarios. One is produced as a standard food-grade product, which needs less energy and emits less carbon. The other assumes that lab-grown meat must be produced using biopharmaceutical processes; this means it needs a very energy-demanding purifying stage to remove contaminants called endotoxins. If it was produced in this way, it could have very high emissions.

Biotechnology experts pushed back. This energy-demanding stage of the process probably won't be needed. Lab-grown meat will be safe and contaminant-free without it (there will, of course, need to be strict food agency standards to make sure this is the case). So, while we will need to wait until commercial lab-grown meat hits the shelves to know its final carbon footprint for sure, it's probably going to be a lot better than the beefburgers people tuck into today.

QUESTION 43

How can people switch when meat substitutes cost more than meat?

Answer: Meat substitutes are too expensive, but there is an opportunity to change this.

It's true that the majority of people are not going to pay more for an Impossible, Beyond Meat, or other plant-based burger. Many can't, even if they wanted to.

I know about these costs after years of buying meat substitutes myself. But anecdotes don't fly here, so I had to run the numbers too. I studied retail prices from national supermarket data in the UK. My focus was on meat substitute products, not the "classic" plant-based proteins like peas, beans, and lentils, which *are* cheaper.

You can see the results in figure 43.1. In every category, meat products stole the cheapest spot. Some supermarket-branded substitute burgers and sausages weren't much more expensive than the meat version. But for bacon and chicken, the cheapest substitute product was more than double the cost of the cheapest meat.

This is not just the case in the UK. In the US, studies find that meat substitutes are 25% to 115% more expensive depending on the type and meat quality. The average price premium is half to two-thirds. So, if chicken cost $4 per pound, the substitute version would cost around $6.50. Quite a difference.

Now, we can argue that rather than meat substitutes being too expensive, meat is too cheap. Meat's price tag doesn't reflect the "true" cost if we were to include its environmental damage and animal welfare costs. Many countries

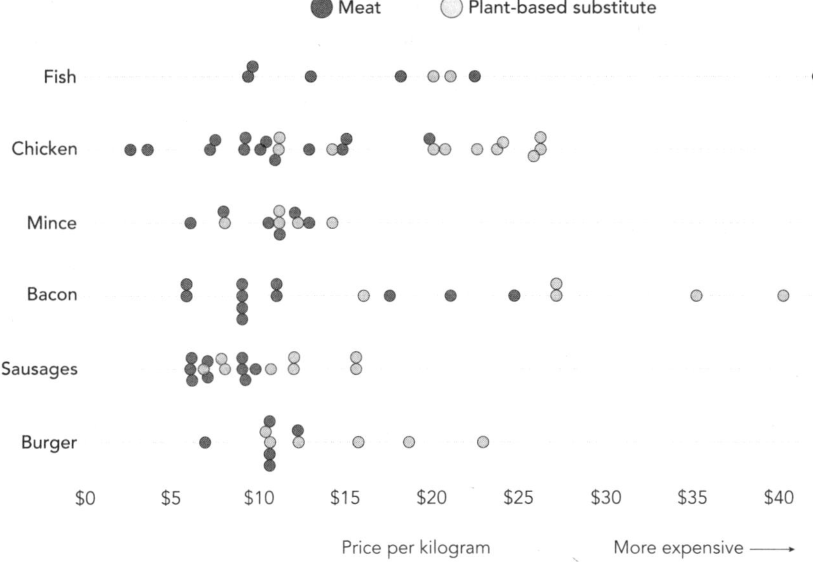

Figure 43.1

The cost of meat substitutes versus meat and fish. Based on retail prices in UK supermarkets. Measured in dollars per kilogram (converted from pounds sterling using a conversion factor of 1.3).

also give their livestock sector large subsidies, which brings the price down for consumers (although someone is ultimately paying for it through taxes). This is particularly true in Europe, where around 80% of the EU's Common Agricultural Policy subsidies are given to animal products.[1] There are obvious solutions to rebalancing the economics (see "What we need to do") but none of these options will be politically popular.

If making meat more expensive isn't on the cards, is there hope that meat substitutes could get cheaper? Perhaps. It's still a young industry, so there is plenty of room for innovation and optimization. As products scale, there might be ways to reduce costs.

WHAT WE NEED TO DO

We can make meat substitutes more cost-competitive with meat in two ways: make meat substitutes cheaper or make meat more expensive. The first solution would be much easier politically (unsurprisingly, people don't like things getting more expensive). From a nutritional and socioeconomic perspective, I'd also be concerned about making protein more expensive: There are already 3 billion people in the world who can't afford a complete, healthy diet. Making food more expensive is not going to help. I want to see cheap *and* nutritious food. What options do we have?

- **Buy meat-substitute products.** One of the best things you can do to bring down the cost of meat substitutes *as an individual* is to buy these products today. That sends a signal to the market—to existing companies, investors, and future entrepreneurs—that they're in high demand and there are people to serve. In the short-term you might imagine that high demand means higher prices. But it makes markets more competitive, and companies will have to compete on price.
- **Put a tax on meat and dairy products (maybe).** Let's set my personal view aside and consider a meat tax. The purpose of the meat tax is to capture the true environmental and social costs of producing meat. Political feasibility is the biggest barrier; no country has ever done it. That seems odd because it already happens in energy systems: Countries put a carbon price on fuel or provide financial support for early renewables. But we don't think about food in the same way as energy. We want the freedom to eat what we like, and making meat more expensive is a step too far for many.
- Surveys in the UK find that around one in five people say they would support a meat tax.[2] That's too low for a politician to feel confident in imposing it. Interestingly, support for a tax on *processed* meat was higher. There's also a case for doing this on the grounds of health, but the UK's 2021 National Food Strategy found that a meat tax was the least popular of any of the suggested interventions: "*The idea of introducing a 'meat tax' was a non-starter. Every time we raised it, the atmosphere would suddenly crackle with hostility. Although a minority of our panelists liked the idea, many more were vehemently opposed—and the arguments between these instantaneous tribes were fierce.*"[3]

Surveys in the US suggest that support for a meat tax is also a minority position. One-quarter to one-third of Americans say they would support one, but this was for a relatively small tax of just 10%. In those same surveys, two-thirds explicitly said they would *not* support a tax, so I expect there would be a lot of pushback. Good luck to any politician trying to make this a reality.

- **Reallocate subsidies for the meat and dairy sector.** Rather than add a tax for meat, you could change the amount of agricultural subsidies farmers and manufacturers receive. The impact on consumer prices might be similar, but reducing a subsidy has a slightly different vibe than slapping on a tax. Reallocating these subsidies from meat and dairy to more climate-friendly substitutes would help develop the meat substitute sector and bring the costs down.[4]
- **Give regulatory approval to lab-grown meats.** Companies can't try to scale lab-grown meat if countries won't allow it to be sold there. In most countries in the world, it's still illegal to sell lab-grown meat because it doesn't have regulatory approval. Singapore and the US are the exceptions (plus the UK where it's allowed in pet food). Some countries are actively pushing back, with Italy and some US states completely banning it. If we want it to have a fair shot at competing—and if we want to give consumers freedom to make their own choices—then these products need to have the same seller's rights as meat. Without it, the industry will be impossible to scale.

THINGS TO BEAR IN MIND

There are also new technologies that could hit the market. Precision fermentation is an area with lots of potential. This technology taps into microorganisms to produce alternative proteins, although you can build molecules such as fats, vitamins, and flavors too.[5] The defining factor of precision fermentation is that it produces more from less, so with the right strains of microorganisms and some clever engineering, it might be possible to produce large amounts of protein very cheaply.

And, of course, there is still hope for lab-grown meat. Its costs have plummeted over the last decade, in a similar fashion to solar, wind, and batteries. Producing a kilogram of cultivated meat in 2013 cost around $2.4 million. That's one expensive burger. Today, the costs are about 50,000 times lower: around $10 to $100 per kilogram. Every couple of years, the cost of lab-grown meat has fallen dramatically. It still has some way to go to be competitive with chicken, but it is starting to close in on the more expensive meats like beef.

If its technology curves continue to follow in the footsteps of solar or batteries, producers think it can get even cheaper. Then the real challenge is scaling these processes up so products can hit supermarket shelves. The next decade could see affordable lab-grown meat breaking through. If it can compete on price, it might be the game-changer we need to give people want they want, at a price that they and the planet can afford.

QUESTION 44

Meat substitutes are "ultra-processed"; doesn't that make them unhealthy?

Answer: Meat substitutes can be just as nutritious as meat, and can be enjoyed as part of a balanced diet despite being ultra-processed.

If I got a dollar every time I heard someone say that meat substitutes are ultra-processed junk food, I'd be able to subsidize the entire industry myself. But how unhealthy are they, really? Let's take a look.

Meat substitutes tend to be lower in calories, total fat, and saturated fat than processed meat.[1] This is good for most of the population because we know that being overweight can increase the risk of health conditions such as heart disease, stroke, and cancer. They're also higher in fiber, which is another positive.

The main downside is that they often have less protein. While some products have the same amount of protein as a beefburger—around 20 to 25 grams—a lot fall short and only provide around 15 grams. There is also a lot less consistency across products: Some will have 20 grams, others 15 grams, and some have even less.[2] That means you have to be pretty vigilant about checking the nutrition labels if you want to guarantee that you're getting a good portion of protein.

The *quality* of the protein also matters. We want to get a full spectrum of amino acids, which are the building blocks of protein. Animal products do have a complete range of amino acids, but individual plant products are often missing one or two. We *can* get the full spectrum by having a varied plant-based diet, but this takes a bit more effort than just eating a beefburger and putting milk on your cornflakes. With the right balance of foods, plant proteins can be just as good for building and maintaining muscle and strength.[3]

What about micronutrients, the essential vitamins and minerals that we need to maintain good health? Meat substitutes in their "raw" form often fall short on a few key ones, like calcium, zinc, and, especially, vitamin B_{12}, which is *only* naturally available in animal products. But many products are now fortified with these nutrients, meaning manufacturers add them in during processing.

But now we come to the juicy and controversial bit: food processing. Quite literally the juicy bit, because heme and other ingredients are added to some meat substitutes to give them the texture of real meat. Meat substitutes for burgers, sausages, bacon, chicken, and mince are all considered "ultra-processed" on typical food ratings. I say "typical food ratings," but there is no global standard for what counts as an ultra-processed food. Most of the food we eat today is processed in some way; it's the extent that varies.

The problem with the rise in popularity of blanket labels like "ultra-processed" is that they suggest all "ultra-processed" foods are equal. Some celebrity doctors would have you convinced that supermarket bread is just as bad for you as cakes and Coca-Cola. This is obviously not the case, and it's unhelpful to throw all these foods into the same basket.

Yes, there is strong evidence that people who consume a lot of ultra-processed foods tend to have worse health outcomes. They're more likely to be overweight, suffer from diabetes, and are at higher risk of cancer and cardiovascular disease. But studies have shown that the food groups associated with clear negative health outcomes are processed meats and sugary drinks.[4] Other groups—including meat substitutes—have no impact on disease risk, and some ultra-processed foods have even been associated with a *lower* risk.[5] It's wrong to automatically assume that more processed means more unhealthy.[6] There's no evidence for that.

Meat substitutes hit some of the criteria that are often used to define ultra-processed foods: They're convenient, contain multiple ingredients, and "couldn't be made in a normal kitchen" (whatever that's supposed to mean). But there are several criteria that they *don't* meet: They're usually not high in

Criteria	Processed meat	Plant-based meat
Convenient	✓	✓
High in calories	✓	
High in sugar		
High in saturated fat	✓	
High in salt	✓	
Many ingredients	✓	✓
Low fiber	✓	
Couldn't be made in "normal" kitchen	✓	✓

Figure 44.1

Conventional processed meat and plant-based meat substitutes measured against the criteria for "ultra-processed food." Based on an assessment of the NOVA processed food classification scheme.

Data source: Good Food Institute, 2023.

calories, sugar, or saturated fat, nor are they low in fiber. What *does* hit most of these other criteria? Conventional processed meats like sausages, bacon, and burgers. If you're switching out a pork sausage for a plant-based one, you're sliding down the ultra-processed scale and away from some of the elements that might make these foods worse for your health.

If we step back from specific nutritional comparisons and look at people's health overall, there are a various studies that suggest that higher consumption of plant-based proteins is associated with better health outcomes and lower chances of early death.[7] Again, some people take these studies too far and assume that vegan diets come with a health halo. That's not it. You can have a healthy diet with or without animal products, as long as it's balanced. But these studies show that a shift to more plant-based diets is not necessarily *bad* for your health.

No one is going to argue that you should be eating meat substitutes *all the time*. But as part of a diverse and balanced diet? You shouldn't be too concerned.

WHAT WE NEED TO DO

While meat substitutes can be part of a healthy, balanced diet, there are things that manufacturers can do to improve their health profile.

- **Make sure they contain at least 20 grams of protein per portion.** Sorry, but if you're offering only 10 grams of protein, please don't insult the consumer and label it "high protein." It is not.
- **Add essential micronutrients like vitamin B_{12}.** Meat substitute brands should always add this, since it's the "missing link" that's needed to make it comparable to a meat product. Many companies—like Beyond Meat—already do that, but it should be an industry standard so we don't need to check the label.
- **Create a more nuanced discussion around ultra-processed foods.** The scaremongering headlines are not helping, nor are the books demonizing almost every food we eat. No, eating ultra-processed foods all the time is not recommended. Yes, *some* of those foods are associated with poorer health outcomes. But not every food that "couldn't be made in a normal kitchen" is bad, and I wouldn't take nutrition advice from anyone who says it is.

IX

CEMENT, STEEL, AND OTHER "HARD-TO-ABATE" INDUSTRIES

To stop warming, we need to get emissions to zero. Electricity and cars are the easy bit. Finding solutions for cement, aviation, shipping, fertilizers, and plastics is much harder, hence why they're often called "hard-to-abate" industries. But hard does not mean impossible.

We're already making progress in building new solutions. In 2021, just 50% of the solutions we need to decarbonize our entire economies were available and already on the market. The other half were either nonexistent or still under development. By 2023, so in just two years, this had jumped to 64%. That's rapid progress that many people don't see. And even the technologies still "under development" are getting closer and closer to breaking into the mainstream.

Let's look at our options for decarbonizing aviation and shipping, cement, and steel. As it turns out, the prospects for finding cost-competitive solutions gets better with every year that passes. We're not there yet, but these industries have moved from "impossible to fix" to "hard to fix," and in the next few decades could shift again to being solved.

QUESTION 45

Will we ever be able to produce low-carbon cement?

Answer: We now have several solutions to reduce or eliminate CO_2 emissions produced by cement.

Concrete is the second-most used material on Earth, after water. As people escape poverty and billions move to cities, we're going to need a lot more of it to build roads, homes, water and energy networks, and infrastructure.[1]

That's a problem for the climate. Cement emits around 7% of the world's greenhouse gas emissions—more than aviation and shipping combined.[2] The issue with cement is not just that it uses energy; it's that the manufacturing process emits carbon dioxide directly. Every tonne of cement we produce emits around one tonne of CO_2.

We make cement by taking a limestone rock ($CaCO_3$) and heating it to very high temperatures. It forms calcium oxide (CaO), which is used in cement. But it also forms CO_2 as a by-product. There's no way to avoid that. Currently, that CO_2 is just released into the atmosphere where it warms the climate.

This is the formula we need to fix:

$$CaCO_3 \rightarrow CaO + CO_2$$

We do have solutions. Not just one but several. They are shown in figure 45.1.

First, we could *reduce* the amount of cement that's used in concrete. People often use "concrete" and "cement" interchangeably, but they're different. Cement is an ingredient. Mix it with water, sand, and other materials, and

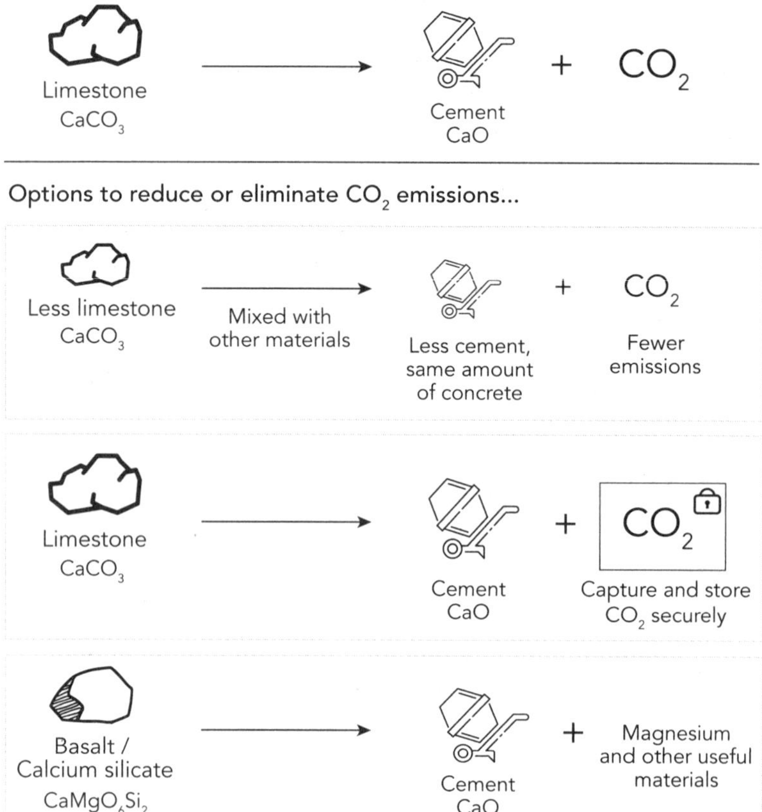

Figure 45.1

Options for reducing carbon emissions from cement.

you get concrete. We could reduce emissions by making concrete that has less cement in it. We know this concrete is sturdy and works because many countries *already* use different amounts of cement. Construction industries in Europe and North America tend to use more than other countries. Europe currently uses a mix that requires around 78%. Reducing this to 50% by 2050 would slash the industry's CO_2 emissions by 40%.[3]

This could help us *reduce* emissions, but it won't eliminate them completely. If we want to do that, we have two other options.

We could capture the CO_2 before it's lost to the atmosphere using carbon capture and storage (CCS), which I discussed in Question 7. The captured CO_2 would have to be stored securely under the cement site or in a depleted oil field. It's technically possible to do this: A company called Heidelberg Materials is already building the first industrial-scale plant in Norway. The challenge will be cost. The reality of CCS is that even if we can get the price down, it's *always* going to cost more than producing cement without it.

Bill Gates often talks about the "green premium": how much more expensive a product is for the low-carbon version. In his 2021 book *How to Avoid a Climate Disaster*, he put the premium for cement with CCS at 75% to 140%.[4] The average price of cement was $125 per tonne. For CCS, you'd need to pay an extra $100 to $200. No one wants to pay two or three times as much for cement. The economics are improving, though. Some industry experts estimate that in places with the cheapest energy, you would "only" need an extra $40 per tonne.[5]

But what if you could make cement without producing any CO_2 in the first place? That's our third option: using a different source rock. Rather than using limestone, where CO_2 is unavoidable, we could use a rock rich in calcium silicate, like basalt. It doesn't contain carbon, so there isn't any CO_2 when it's heated at high temperatures. Instead, the process produces *useful* by-products: "supplementary cementitious materials" that can be used in concrete, and magnesium compounds that can absorb CO_2 from the atmosphere. The other good news is that basalt is one of the most abundant surface rock on Earth, so running out will not be a problem.

Over the next decade, I think several solutions to low-carbon cement will enter the market. It'll then be a tussle to see who can make it as cheaply as cement today. This is a dramatic change from 10 years ago. With proper backing and incentives, incredible scientists are showing that it might not be the impossible problem we once thought.

WHAT WE NEED TO DO

We have at least three potential solutions to decarbonize cement. For each, there is a clear barrier that we need to push past.

- **Reform construction regulations to allow concrete with less cement.** Companies could cut emissions tomorrow if regulations across the world let them build with concrete that had a lower concentration of cement. We know this works because some countries are already doing it (and the buildings are still standing). Of course, we do need to ensure strict safety standards, but let's put it to the test, and then get building.
- **Governments can provide economic incentives for carbon capture and storage.** CCS will never be cheap, and current cement companies are not going to spend on capturing CO_2 unless they're forced to. It doesn't make sense for their bottom line. But governments can encourage this by offering tax reductions or incentives for every tonne of CO_2 that's captured. This already happened in the United States: In its Inflation Reduction Act, the US offers a tax reduction of $85 per ton for CCS if stored permanently.
- **Invest in innovation and early-stage start-ups that are building new solutions.** A decade ago, most of us thought that CCS was going to be our only (and expensive) option for cement. But innovators are proving us wrong. Solutions that use a different process entirely are the most exciting to me because they could really be cheaper and better than what we have today. One company, Brimstone, has already shown proof of concept for using calcium silicate and is now building a pilot plant. We should be supporting more innovators like this who are trying to break the mold, whether it's through government grants or private investment—ideally, both at different stages in the journey.

THINGS TO BEAR IN MIND

One interesting thing to consider is how the economics of cement with CCS affect the price of consumer products. It's less than you might think. While cement with CCS might cost as much as 75% more to make, the premium on products that *use* cement might be much smaller. I ran the numbers for a house. Assuming that low-carbon cement is 75% more expensive, it increases the cost of a $250,000 house by less than 1%. This is because cement makes up a relatively small percentage of the final cost of a house. It might cost the homebuyer a few thousand dollars more. That sounds like a lot (and it is for some people, especially in lower-income countries), but house prices can swing up and down by far more than a few grand in the space of a month. As a prospective house buyer, I wish an increase in house prices of a few thousand pounds was seen as a lot, but while a 75% cost premium on cement sounds alarming, the impact on final products might be so small that many people don't notice.

QUESTION 46

What about low-carbon steel?

Answer: We know how to produce green steel; we now need to build it at scale.

Steel isn't something we think about much, but it emits around 7% of the world's greenhouse gas emissions.[1] Like cement, the world will need a lot more steel in the next 50 years to build cities, bridges, and infrastructure for billions, not to mention the wind turbines and other low-carbon technologies we'll need to tackle climate change.

For a long time, low-carbon steel seemed impossible. But we now have a couple of options. The final hurdle is getting the economics to work.

To understand how we decarbonize steel, we need to understand a little bit about the process. Two-thirds of the world's steel is made from coke, a product of coal.[2] One tonne of new steel needs around 0.8 tonnes of coal. That's where the CO_2 emissions come from. The remaining third of steel is produced from electric arc furnaces. This takes scrap steel and recycles it back into new steel. This has lower carbon emissions, especially if the electricity comes from low-carbon sources, but there isn't enough scrap metal to produce *all* our new steel from electric arc furnaces.

There are several ways to decarbonize the process of making new steel. We can capture the CO_2 that's emitted, using CCS. This would work but is too expensive.

The alternative is to get rid of the coal in the process. First, by using hydrogen to strip the oxygen from the iron ore. Instead of producing CO_2, you produce steam. Second, by directly electrifying the process. Which route proves to be the most promising is still an open question. Both options will need a lot of electricity, which is the main challenge.

Producing one tonne of "green steel" would need 4 MWh of electricity.[3] If the world produces 1.3 billion tonnes of new steel each year, that adds up to 5,200 TWh. For context, that's equal to around one-fifth of global electricity generation today. So "green steel" would increase global electricity demand by a quarter. That's a lot.

How much would it cost? We can use the numbers from a company already building its first large plant: H_2 Green Steel, a plant in the Baltics.[4] If we take its investment figures and scale it up to cover *all* the world's steel production, it would cost €1.3 trillion. If you split this over 25 years, then it would cost €50 billion per year. Again, not small numbers but not totally unfeasible, especially with government support.

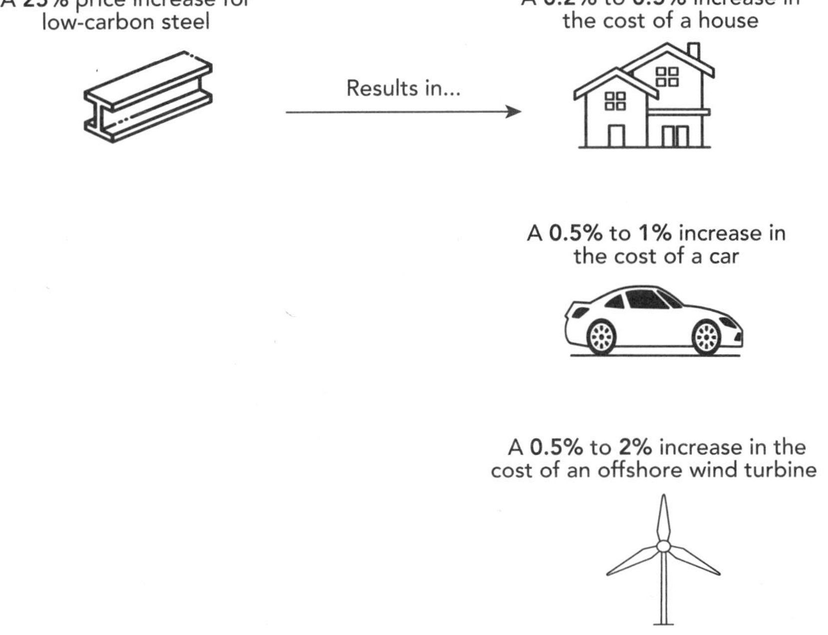

Figure 46.1
What difference would a green steel premium make to the price of already expensive products?
Data sources: Author calculations based on ETC, BNEF, and RMI.

If manufacturers can get investment to start building green steel, there will be no shortage of markets for it. Many companies that use steel in their products—car makers, construction companies, household appliance manufacturers—will want to say that the steel they use in their products is low-carbon. Since steelmaking emits so much carbon, cleaning up that bit of their supply chain will make a massive difference to the footprint of their own products.

Even if green steel costs 25% more, it might not make much of a difference to the final price of products.[5] Look at the examples in the chart. A car would only be around $150 more expensive, which is barely noticeable for a $25,000 car. For a house, this "added cost" would be around $400. If a home costs $250,000, that's an increase of just 0.2%. Again, barely noticeable.

If steelmakers can keep the costs manageable, companies will be clambering to get their hands on the stuff. Not long ago it was the question of whether you *could* produce green steel. Now it's just a question of when.

WHAT WE NEED TO DO

We are on the cusp of several green steel solutions breaking through. But decarbonizing all the world's steel is going to be a massive task. We need to build quickly, and on a vast scale. Here are some things we can do to go faster.

- **Build steel plants where electricity is cheap.** Our low-carbon options will need a lot of electricity, so electricity needs to be cheap to keep the costs down. The problem is that many steel plants are either near coal mines or in countries with relatively expensive electricity.[6] We need to rethink where we build new plants to keep prices down.
- **Hold manufacturers to their commitments.** Most European steel groups have made commitments to reduce their emissions by 30–50% or more by 2030. If they're to meet these goals, they'll need to start building this infrastructure now. Governments and companies further down the supply chain should be holding them to this.
- **Governments can provide financial support for early innovators.** This is already happening in a few countries. The Swedish government is

supporting the H_2 Green Steel plant. The Spanish government is doing the same for ArcelorMittal, which is converting an existing steel plant to run on hydrogen. But outside of Europe, they're lagging behind. Other markets need to get a move on.

- **Private companies can provide early investments.** Boston Metal is one steel manufacturer that has backing from several large companies, including Microsoft, to launch their first plant by 2026. We need more of these partnerships to deliver green steel quicker, and at scale.

QUESTION 47

Is there any hope of low-carbon aviation and shipping?

Answer: Flying and shipping can both go low-carbon, but getting the costs down will be the main challenge.

Aviation is the Achilles' heel of so many climate warriors (including me). They've given up meat, they don't have a car or have gone electric, and they've installed a heat pump. But flying is the one area they struggle to cut out completely. While people might reduce how much they fly, the reality is that many people are not going to stop it. Frequent fliers could rein in on the weekend hops to Europe every month. Many in business could dial in remotely for some—even if it's not all—of their meetings. But some people fly to visit relatives living on another continent, and they'll be getting on that flight regardless. Others are not going to tell their boss that they will *never* fly for a business trip.

Do we have any chance, then, of flying in a low-carbon way? And can we do the same for shipping? Each of these sectors emits around 2–3% of the world's carbon emissions. Aviation contributes a bit extra to global warming because of "contrails"—the trails of water vapor that you often see behind planes—because water vapor causes more warming at altitude.[1]

Neither sector is making fast progress. And that's me being polite. To be fair to these industries, decarbonizing is a *hard* problem for them. Building wind farms is easy compared to sending an 80-tonne container 30,000 feet into the atmosphere in a sustainable way. But what's also slowing progress is that no country is responsible for international aviation or shipping. That means they can simply ignore it and focus on the emissions that make a difference to their report card. While many airlines have committed to flying "net zero" by 2050, it has been mostly talk and not a lot of action.[2]

It is technically feasible for aviation and shipping to go low-carbon, but what's lacking is the investment to make it a reality, time to bring the costs down, and trade-offs with other problems like land use. Let's briefly look at the options for each.

Shipping emissions have been creeping up, but fairly slowly.[3] One reason to be optimistic about reducing emissions is that a lot of the *demand* for shipping will disappear in a low-carbon world. Forty percent of the stuff that's shipped around is fossil fuels.[4] Get rid of fossil fuels, and global shipping demand drops a lot.

What can we do about the rest? Ferries and other short journeys could go electric. Norway already has more than 60 electrified ferries running.[5] These have yet to break the 100-kilometer barrier, though, so unless we see dramatic improvements in battery technologies (which is possible), we will probably need another technology for longer journeys.

There are currently three main fuels in the running: methanol, ammonia, and hydrogen.[6] The challenge for each will be cost. The current price of "green" methanol is around $320–720 per tonne.[7] It would need to get down to at least $250 per tonne to compete with fossil fuels. It's early days, though, and there is plenty of room to reduce costs and improve efficiency.[8]

What about aviation? Its emissions have doubled since 1990.[9] Looking forward, even in a scenario where people cut back on flying, we're going to need a lot of sustainable fuels. Much like shipping, some shorter flights could be electrified and run on low-carbon electricity. For long-haul journeys, it could be some combination of biofuels, hydrogen fuel cells, or synthetic fuels from hydrogen. Each has its downsides.

For biofuels, it's mostly about land use and the risk of food insecurity. In 2023, Virgin became the first airline to fly across the Atlantic Ocean using waste fats as the main fuel. Now, the world is never going to have enough waste fats to power the aviation industry, but it *could* use vegetable oils instead. The problem is that the world produces around 200 million tonnes of vegetable oils each year, and aviation would need 250 to 300 million

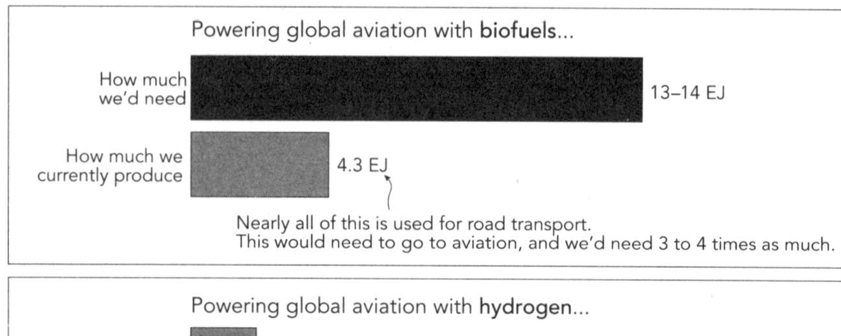

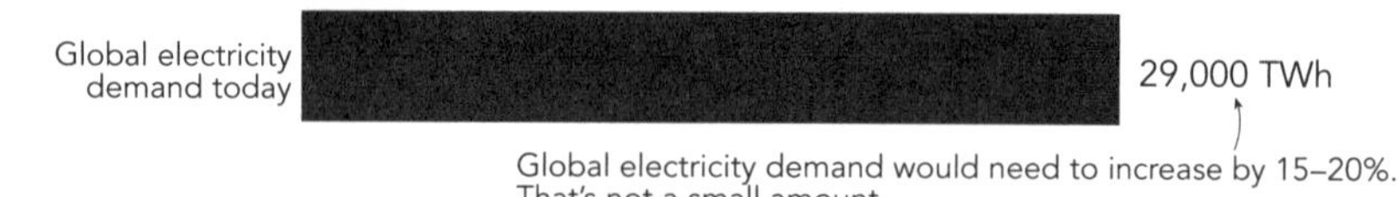

Figure 47.1

Decarbonizing aviation using biofuels or hydrogen. The challenge for synthetic fuels like hydrogen is the amount of electricity required to produce the hydrogen in the first place. To power the aviation industry today would need around 4,000 TWh. That's 15–20% of global electricity.

tonnes. That means global vegetable oil production would need to more than double to feed people and power our airplanes. Maybe we could use cereals for biofuels instead. That's still a massive challenge. The world would need to use *all* of its maize production (which is used for food and feeding animals, so is unavailable) just for aviation.[10] Or three to four times the amount that's currently used for biofuels for road transport. To put this in context, growing that amount of cereal needs an area the size of India.

Neither hydrogen nor biofuels is impossible. But the world will need to make a choice and accept some of the trade-offs that come with it. The most obvious hurdle, in the end, is going to be cost. Fuel makes up around one-third of the ticket price of your flight, so consumers are going to feel it.

Aviation and shipping will be two of the last sectors to decarbonize, and people are right to be skeptical that we can ever pull it off. But the world does have options; it's about figuring out which ones could get us there soonest and at the lowest cost.

WHAT WE NEED TO DO

Decarbonizing shipping and aviation won't be easy, but a combination of behavioral, economic, and technological interventions can help us first reduce then get rid of travel emissions completely.

- **Cut back on flying (although we'll never eliminate it).** The world is not going to stop flying, but the less we do, the easier it is to decarbonize. If you fly often, thinking about how you can fly more selectively would help a lot. For companies: Go virtual when you can, and save in-person events for important gatherings.
- **Put some of the costs on frequent fliers and private jet-setters.** Consumers are probably going to have to foot the bill for low-carbon flying, but one way to protect consumers who rarely fly is to implement a frequent flier tax, where those who fly and pollute the most cover a lot of the costs. Studies show that these taxes could raise considerable amounts of money, with nearly all of it coming from the top 20% of fliers in each country. Private jet-setters should (must) also play a role: They could lead the way by investing in electrification for short journeys, and alternative fuels for long ones.
- **Put more pressure on the aviation industry to deliver.** Since no country takes responsibility for the emissions from international travel, the aviation and shipping industries have had an easy time of it at international climate negotiations. They mostly get a free ride. Decarbonization is not going to be cheap for them, so they won't do it unless the pressure is on. It's time to start.
- **Keep iterating and innovating on battery technologies.** I've underestimated the progress on batteries in the past, and despite being a proud optimist, I'll probably be surprised by their progress in the future. A lot of shorter boat and plane journeys will be electrified. How far they'll go is still an open question.
- **Experiment with alternative technologies for longer journeys.** There isn't a clear winner yet, so it'll take companies investing in different solutions to find the best one. Maersk, one of the shipping industry's leaders, has

already started building methanol-fueled vessels, with 18 expected to come to market by the end of 2025.[11]

THINGS TO BEAR IN MIND

One other option that companies have to decarbonize aviation is for airlines to keep using jet fuel, and then pay to suck CO_2 out of the atmosphere using carbon removal methods (see Part X). It's the most controversial option but could end up being the cheapest. If the costs of technologies like direct air capture could get as low as $100 per tonne (which is a big "if"), then fuel costs would increase by around 50%. That's less than the cost hike to use hydrogen today, although that could change in the future if hydrogen gets cheaper. This is not too different from the current concept of "carbon offsets," except we'd want to do it properly. We'd have to guarantee that the CO_2 being sucked out of the atmosphere was stored securely over long periods of time; that's not the case with offsets today, many of which are a scam.

X

CARBON REMOVAL AND SOLAR GEOENGINEERING

Humans are tinkering with Earth's thermostat by increasing the amount of greenhouse gases in the atmosphere. What if we can change this by actively sucking carbon *out* of the atmosphere or deliberately cooling the planet down?

The first of these interventions is called carbon removal. Here I'm *not* talking about carbon capture and storage. CCS is when we capture the CO_2 that's emitted to stop it from going into the atmosphere in the first place. With carbon removal, we're actively removing CO_2 that's *already in the atmosphere*. CCS is trying to stop the mess; carbon removal is trying to clean it up.

The second is called "solar geoengineering," which aims to cool the planet by reducing the amount of sunlight that reaches Earth. There are a range of technologies that, in principle, could work. We could add tiny reflective particles into the upper atmosphere where they reflect some light back to space; we could spray salt from seawater into the atmosphere to brighten marine clouds; or put large mirrors in space to shield some light from the sun (which is now mostly a science-fiction proposal).

Many people are skeptical about both sets of technologies. Should they play a role, and how much are they going to cost us?

QUESTION 48

Can we solve climate change without carbon dioxide removal?

Answer: We need to cut out most of our emissions completely, but there might be 10–15% that we'll need carbon removal to help with.

There are almost no future pathways from climate and energy experts that get our emissions to zero by 2050. There's always a bit left over, because of industries like plastics, aviation, and agriculture, which will struggle to switch to low-carbon sources or to using CCS.

Carbon removal is not going to solve climate change, but we might not be able to solve climate change *without* it. Reducing our emissions should always be top priority. I'm not talking about reducing emissions by 30% or 40%; we need to cut them by more than 80%.

Scenarios from the Intergovernmental Panel on Climate Change tend to assume that we can cut our emissions by around 85% using current and future technologies, meaning 6 billion tonnes of CO_2 per year would be left for carbon removal to handle. Other scenarios range from around 2 to 10 billion tonnes.[1] It depends on how successfully we tackle difficult sectors like plastics and agriculture.

We are nowhere close to capturing billions of tonnes today. Despite such groundbreaking innovations as direct air capture, they are mostly prototypes. The world's direct air capture plants are removing around 10,000 tonnes per year. That might sound a lot, but it's just 0.0003% of our current emissions. What about the "natural" way of removing carbon: planting trees? Well, yes, we should be planting a lot more trees and restoring ecosystems. If we can transform our food systems, freeing up billion of hectares of farmland (see Question

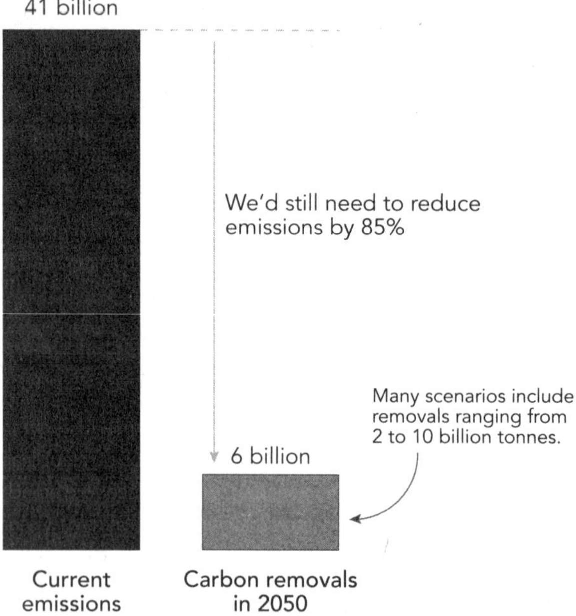

Figure 48.1

The amount of carbon removal we *might* need by 2050.

Data source: Current emissions data from the Global Carbon Project. Carbon removals based on a review of the scenarios literature. The amount of removal needed varies but is usually in the billions of tonnes.

41), we'll have a lot more room to do this. But as the climate scientist David Ho notes: "If everyone on Earth planted a tree—8 billion trees—it would take us back in time by about 43 hours every year, once the trees had matured."[2]

The current technological and natural solutions are both very far away from what we need. So we will need to find a way to scale these solutions, and we have a few decades to work it out. Twenty-five years ago, solar power produced less than 0.01% of global electricity. I'm glad the world didn't give up on it then.

WHAT WE NEED TO DO

If we want cost-effective solutions to get to net-zero emissions, here's what we need to do.

- **Reduce emissions, reduce emissions, reduce emissions.** We're not going to "carbon removal" our way out of this mess. We need to drastically cut emissions, then work on carbon removal for the small slice at the end.
- **Early investment into carbon removal, and support from governments and private companies.** That doesn't mean committing 80% of our climate spending to removal technologies. It doesn't even mean 50%. Most funding should be going into solutions that we know will reduce emissions today: clean energy, electric vehicles, heat pumps, and innovations that can decarbonize cement and steel. This is already the case. In 2023, the removal industry got $1.2 billion of early-stage investment—a tiny fraction of the $1.8 *trillion* spent on clean energy as a whole.[3]
- **Set up early market commitments for carbon removals.** Working on new technologies that can do carbon removal is risky: It's expensive to fund, and there's no guarantee that there's going to be demand for your product if you're successful. "Advanced market commitments" try to solve this by getting companies to commit to buying carbon removals far in advance so that entrepreneurs and investors know that there is a guaranteed market, and they can get good returns. This is already happening through programs such as Frontier, founded by companies like Stripe, Shopify, and Alphabet, who have committed to at least $1 billion of removals.

QUESTION 49

Isn't carbon dioxide removal too expensive?

Answer: Carbon removal technologies are expensive today, but if we can get their costs down, it might be the most cost-effective way to tackle emissions in some sectors.

There are a bunch of innovations that could help remove carbon from the atmosphere, including restoring ecosystems, biochar, faster weathering of rocks, and marine carbon removal. Though these innovations could play a key role, to keep things simple I'm going to focus on another one—arguably the one that gets the most attention—direct air capture. This uses machines to literally suck CO_2 out of the atmosphere.

Okay, back to the costs. Climeworks—a company in Iceland that is already sucking CO_2 out of the air—has said that it costs them around $1,000 to remove one tonne of CO_2.[1] So, removing the 40 *billion* tonnes of CO_2 we pump into the atmosphere every year will cost around $40 trillion. Climeworks has also quoted figures of $500–600 per tonne, so we might be lucky and get it half-price. Either way, no government or rich philanthropist is going to step in and foot that size of a bill. The hope, though, is that we'll do two things that will help:

1. **Dramatically reduce our emissions.** If we were emitting 4 rather than 40 billion tonnes of CO_2 each year, the price tag would be $4 trillion, not $40 trillion.
2. **Make carbon removal much cheaper.** Many of these technologies (apart from the good old practices like planting trees) are in their early stages, so they're expensive. The unofficial aspiration for the industry is to get these costs down to $100 per tonne. No question, this is incredibly

ambitious, but if they achieved it the price would also shrink from $40 trillion to $4 trillion.

Solar panels, wind turbines, and batteries were crazily expensive a few decades ago. But with sustained investment and deployment, their costs have plummeted. For every doubling of solar capacity, prices have fallen by around 20%. To reach $100 per tonne in 2050, carbon removal technologies such as direct air capture would have to follow a shallower learning curve of 10–15%, meaning every time we double the global capacity of these machines, the cost would have to fall by around 15%.[2] Will this happen in reality? We don't know. Not all technologies follow these beautiful curves. Nuclear energy and fossil fuels certainly haven't. The key is that these technologies need to be modular: the same technology built over and over again. Some carbon removal technologies fall into this category, so there's hope here. Some experts in this area think a drop to $100 to $150 is doable. Others think that costs in the range of $200 to $500 are more likely.[3]

Let's again run some back-of-the-envelope calculations on how much carbon removal *might* cost in the future. Let's say the world does achieve its goal of $100 per tonne—probably our most optimistic cost. And we manage huge cuts in global emissions, pushing them down to just 5 billion tonnes per year. It would cost $500 billion every year to get rid of those final emissions with carbon removal. Again, not cheap, but also not crazy numbers. In 2023, the world invested $1.7 trillion in clean energy, up from $1 trillion less than a decade ago. These investments are only going to rise in the future.

Carbon removal is never going to be a cheap option. But for some select industries, it might be the *cheapest* option to get to net-zero emissions. We will only find out how cheap these technologies can be if we put *some* money into building them today, though not a lot compared to technologies that *reduce* emissions.

WHAT WE NEED TO DO

Regardless of whether we manage to get the costs down to $500, $200, or $100 per tonne, costs need to fall quickly from where they are today. How do we get there?

- **Invest in these technologies early.** Most governments have committed to reaching net-zero emissions in the next 25 years—many by 2050. But none know how to get there with carbon reductions alone, so they're banking on some carbon removal too. If they want these technologies to be ready on time, they need to give them some financial backing today.

  The massive drop in the cost of solar power wasn't inevitable. It was backed early by generous government subsidies in countries such as Germany and China. Without those, the technology might not be where it is today. The same is true for carbon removal. Wait until 2045 or 2050, and it will be too late. As we've seen, some countries are already doing this. Biden's Inflation Reduction Act in the United States offered a subsidy of $180 per tonne for permanent carbon removal and storage, and unlike support for other climate solutions, it looks like this support *might* remain under the new administration. It's not going to cover all the costs of early stage technologies, but it's a start.
- **Companies also need to get serious.** A lot of brands brag about their "ambitious" net-zero plans but also hide that a big chunk of those commitments relies on carbon removals. If they really have parts of their supply chain that are *extremely difficult* to decarbonize, then fine. But they need to put money on the table to make these removal options a reality. If they don't, they are effectively greenwashing and hoping they can piggyback on the investments from others. The worst case—which is the position we're in today—is that too many governments and companies are banking on solutions that don't yet exist and are not willing to invest in getting us there.
- **Reduce energy costs.** Technologies like direct air capture consume a lot of energy. They'll be too expensive if they can't be made more efficient and there isn't a source of cheap electricity to run them. The Climeworks plant in Iceland is already benefiting from cheap and abundant geothermal energy. We'll have to think carefully about where to put these types of plants so that we're minimizing energy costs as much as we can.

QUESTION 50

Isn't solar geoengineering too risky?

Answer: Solar geoengineering is risky, but we need to research it to know *how* risky, as well as its benefits.

This is easily the most controversial topic in the book. On other issues, there is agreement: We certainly need to reduce our carbon emissions, and most experts think we'll need *some* CCS and carbon removal (although they argue about how much).

People really don't agree on solar geoengineering (cooling the planet by changing the amount of sunlight that reaches it). Some think it's now essential to maintain a safe climate, while others think we shouldn't even touch it. My take is that we shouldn't be using geoengineering—at least not at the moment—but should be putting small amounts of funding into research on it.

Many people have argued that we've been running a natural geoengineering experiment for centuries, and not just in terms of global warming. When we burn fossil fuels, we don't only emit CO_2. We also emit local air pollutants that are damaging to our health and ecosystems. But these local air pollutants reflect sunlight back to space and *cool* the planet too. Air pollution has been masking or offsetting some of the warming that we've had from emitting greenhouse gases.

The argument, then, is that we'd be doing the same process with geoengineering, but deliberately rather than accidentally. There is some truth to that, but the experiment is not exactly the same. When we generate local air pollutants, they hang around in the lower atmosphere, close to the ground.

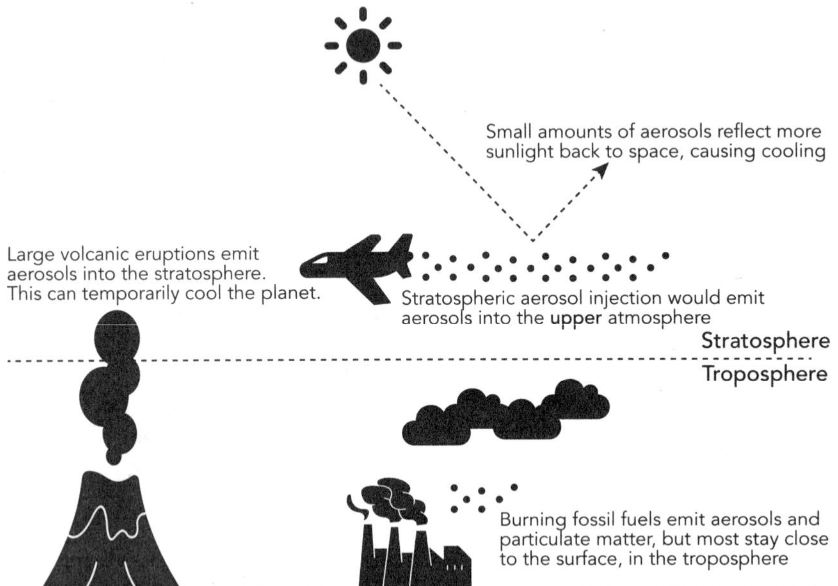

Figure 50.1
Where different types of sun-reflecting particles end up in the atmosphere. Some pollutants from burning fossil fuels do cool the planet, but they typically stay in the lower atmosphere: the "troposphere." Some proposals for solar geoengineering put particles in the upper atmosphere (the "stratosphere"). We haven't done that before, but volcanoes have.

With geoengineering such as stratospheric aerosol injection, we'd be shooting it high into the upper atmosphere (the stratosphere). There, particles have around 100 times the cooling impact as they do in the lower atmosphere. As we haven't deliberately put aerosols this high before, we would be in uncharted territory.

Large volcanic eruptions do spit ash and aerosols into the stratosphere, so we have some data from events such as the 1991 eruption of Mount Pinatubo in the Philippines. Global temperatures dropped by around half a degree. That's the first positive of solar geoengineering: We know that it would work; it *would* cool the planet.

In principle, it'd also be pretty cheap. Some estimate that methods such as stratospheric aerosol injection could cost as little as $2.5 billion per year to cool the planet by 1°C.[1] Others give higher figures of $10–18 billion.[2] Around 0.00001% of the global GDP comes to $18 billion. And much less than the world currently invests in clean energy, which is around $1.8 trillion.

I say that it's cheap "in principle" because the main downside to geo-engineering is that we don't know how it could change other parts of our climate system, like rainfall patterns. If it changes them in unexpected and damaging ways, it would be an extremely expensive bet. Do it irresponsibly, and it could cause more harm than climate change, not less.[3]

There are other significant drawbacks: If we did geoengineering without ditching fossil fuels, we wouldn't reduce local air pollution, we wouldn't stop ocean acidification, and we wouldn't get any of the other benefits of the energy transition that I've covered in this book, like improved energy security, more stable prices, and localized energy production.

So, geoengineering is not a silver bullet. We would still need to move away from fossil fuels and reduce emissions at the same time. To be clear, this is what most geoengineering researchers recommend.

Geoengineering, then, could allow us to cool the planet while we get our act together on reducing emissions. Many scientists on the pro-geoengineering side don't advocate for us using it right now. They want to see more research. I'd like to see this alongside a global agreement not to deploy them at scale. This research is important for two reasons:

1. The global climate system is complex, and we're rolling the dice by turning up the thermostat. A huge climate curve ball is unlikely, but the odds are not zero. Having a "break-the-glass" emergency solution to reduce some of these impacts seems sensible. But to weigh up the pros and cons responsibly, we need to know what the potential impacts of geoengineering might be.[4] That would give us a more informed decision about whether the trade-offs are worth it.
2. A single country or handful of countries might choose the geoengineering route on their own. It might be off the table for tiny island states, but not for medium-to-large countries because the technologies are so cheap to deploy. Inequality lies at the heart of climate change, and that means that countries will have different thresholds at which they would

be happy to take a gamble on geoengineering. The countries that will be hit hardest by climate change might eventually say "no more" and put a stop to it. If this is a possibility, the world had better understand what the impacts might be.

Investing *small* amounts of money into geoengineering research (not deployment) seems like a good idea to me. But it's never going to be the ultimate fix. As the NASA engineer Riley Duren put it: "Geoengineering is not a cure. At best, it's a Band-Aid or tourniquet; at worst, it could be a self-inflicted wound."[5]

WHAT WE NEED TO DO

We can't get sidetracked by the promise of a quick fix to solve climate change. That doesn't exist. But we can get on with the hard work of decarbonizing our economies while also trying to understand the risks and benefits of geoengineering.

- **Reduce greenhouse gas emissions.** Surprise: Geoengineering is no substitute for this. We need to get our act together on that first and foremost.
- **Invest in research to understand the risks better.** Not huge amounts, and far less than we're investing in clean energy and emissions reductions. But small amounts of funding for research into technologies that could possibly destabilize the planet if deployed poorly or save lives if done sensibly, seems like a good pay-off.
- **Make sure lower-income countries are also involved in research and decision-making.** It's the poorest countries in the world that are already suffering the worst consequences of climate change. They need to be involved in the research and negotiations that have the potential to save or destroy their future.[6] If they don't have a seat at the table, they might go for it on their own.
- **Maintain a global moratorium on deploying global geoengineering until we understand these risks.** While we're figuring out the risks, there should be an agreement not to push the button.

CONCLUSION

Getting our emissions to zero—while providing a good life for billions of people—is one of the biggest challenges that humanity faces. It's possible to do it, and there are very few technical constraints in our way, but that doesn't mean it'll be easy.

We need to rethink and rebuild almost everything around us. Our energy systems. How we move around. How we heat or cool our homes. How we design our cities. What food we eat and how we grow it. What we build stuff from. Since humans are resistant to even the smallest of changes, remaking the world around us is going to be difficult. Parts of the transition will seem messy as we work to adapt and upgrade our systems to a new reality. The path will not be linear: There will be bumps in the road that will temporarily slow us down or take us slightly off-track. But the journey we're on is a marathon, not a sprint. Our future will not be determined in a single year or even a four-year presidential term. When setbacks do arrive, we need to be determined to push harder and faster when they pass.

It won't be straightforward, but it will be worth it. This transition is not a sacrifice; it's an opportunity to build a better, fairer, and more sustainable world: one that works for those who are alive today and is compassionate for those who come after us. I hope that these 50 questions and answers have made you more optimistic about the tools, resources, and brainpower we have to make this a reality.

When it comes to solving climate change, there's no cookie-cutter template that everyone, everywhere, can follow. But we have a toolkit of solutions—a kit that gets bigger every year—and countries, regions, and cities will have to pick their own selection. Some countries will get far with solar and batteries alone. Others will need to lean much more heavily on wind. A handful won't have the space for renewables and might need to go all in on nuclear or import electricity from overseas. Many cities could thrive with very few cars. In other areas, electric vehicles will hold the key to decarbonizing transport. What's crucial is that we *do* have a whole set of options that are effective, increasingly affordable, and if implemented thoughtfully, can make the world a much better place. Having this menu of options to choose from is an incredibly recent luxury. Just a decade ago, it was slim pickings. Ten years from now, it will be a banquet.

FIVE QUESTIONS TO HELP YOU SEPARATE FACT FROM FICTION

I hope this is a book you come back to when you see a confusing claim and need a sense-check. But it's impossible for me—or any other researcher—to keep up with the pace and volume of new arguments that hit the headlines.

So, I'll leave you with five tips that I use every day to immunize myself from misleading claims. The next time you're reading an article, scrolling social media, or listening to a commentator on television, think about these five questions:

1. **What are the numbers?** A picture might paint a thousand words, but it doesn't give us accurate statistics. Images capture the emotion or mood of a particular moment in a specific place. But they tell us nothing about the lives of 8 billion people. They tell us nothing about what's happening across 500 million square kilometers of the planet. Don't fall for photographs or anecdotes. An image of solar panels in a field tells you nothing about how much land solar power uses. An anecdote about one electric car fire is meaningless in a world with around 1.3 billion cars in it. Combining pictures and stories with data can work well. But those

who *exclusively* make claims based on pictures are playing to emotions, not rationality. Resist the emotive tug and be skeptical if there are no numbers alongside it.

2. **Is that a big number?** Ask this question whenever you see a number in the news. We humans are terrible at understanding scale. Our brain breaks once numbers get into the thousands, millions, and billions. Anything in that range sounds big.

   Take the example of the world's carbon emissions. An intervention that reduces emissions by a million tonnes might sound impressive: Tonnes sound big, and a million is a huge number. But the world emits around 40 *billion* tonnes of carbon dioxide every year. This grand victory, then, reduces emissions by just 0.003%.

   This "big number" fallacy is often used to undermine low-carbon solutions. We're told how many millions of tonnes of minerals we'll need; how many thousands of square kilometers of land; how many trillions of dollars will need to be invested. These numbers seem huge and unsustainable. At least until you put them into context of whether they're big or not. Yes, we'll need to invest trillions into low-carbon energy, but we've been investing trillions into fossil fuels for decades too.

   Don't immediately assume that thousands, millions, or billions is a lot. Try to find context to put numbers in perspective.

3. **Compared to what?** This is a good follow-on from the "big number" fallacy. People often compare the impacts of low-carbon energy to a world without any energy at all. They compare the land used, minerals mined, or waste generated compared to zero. But that's the wrong baseline. The world is going to need energy, so we need to find the least impactful way of producing it.

   The right baseline is our existing system of fossil fuels, or other low-carbon alternatives we might have. Does wind power produce less waste than fossil fuels? If so, that's a win, even if it's not zero. Does solar power need less mining than coal? If so, that's a win. Clickbait headlines often quote the millions or billions of tonnes of the impact of low-carbon energy but fail to mention the number for fossil fuels. Make sure you find that comparison before you dismiss these alternatives.

4. **How old is the data?** The cost of solar power has fallen by more than 90% in the last decade. Batteries too. The cost of wind has dropped by 70%. Renewables and electric cars are taking off quickly. This year, around half of new cars in China will be electric. In 2020, this figure was just 6%.

   This means that if someone is quoting numbers from 2020—which might not seem like long ago—they're completely out of date. It's even worse if they're rhyming off numbers from 2015 or earlier (which is not uncommon). Commentators will often use old data to make it seem like progress isn't happening or isn't technically possible. When you come across quoted numbers, always question whether the data is recent or whether it's from a world and situation that now belongs in the history books.
5. **How could things change in the future?** Are you the same person you were 10 years ago? 20 years ago? Probably not. Most of us would admit that we've grown as a person. Do you think you'll change much in the next 10 or 20 years? Most people also say probably not: They'll be pretty much the same as they are today. This is the so-called "end-of-history illusion" where people admit to personal development in their past but don't think they'll grow in the future.

   This doesn't just apply to individuals. We have the same bias when it comes to societal or technological change. We accept that changes have happened in the past but are skeptical that tomorrow, next year or the next decade will be much different. When it comes to climate technologies, this couldn't be further from the truth. Things are changing quickly. That means the impact of these solutions will only keep getting better. The solutions we have today are as bad as they'll ever be, and that's a good thing. Their impact is only going to improve, and we need to keep this in mind when we're thinking about claims against them.

TIME TO TURN CLEARER THINKING INTO REAL ACTION

Of course, what matters in the end are outcomes, not thoughts or intentions. I didn't write this book so we could sit around all day talking and thinking about what we should do. I want us to use good information to act.

But effective action starts with clear thinking. Wandering around in the dark—fumbling for the lights—is not going to get us there. Rose-tinted glasses won't do it either. We need an objective and honest vision of what options we have, and what opportunities and challenges they create. We need an informed public that can ask governments for the right things. And an electorate that supports them when they follow through (or votes them out when they don't). Individuals who know what changes make a difference in their own lives and can inspire those around them to do the same. Clear thinking combined with determined optimism creates a culture of inspiration, the antidote to cynicism that we sorely need.

I might live and breathe data, but I'm not deluded into thinking that just showing people facts automatically leads to better decisions. Data is necessary for effective change, but it's not sufficient on its own. The real magic comes when smart people combine data with a compelling narrative. Winning narratives often pull at our hearts more than our heads, so we need both: an emotional drive to act, and a logical head to know *how* to act.

To see the world objectively you need to step *back* to take in the full picture. Data provides that 360-degree panoramic view. But we're often moved to change or act by leaning *in*. We're motivated by the people, communities, and memories closest to us. The nonprofit Potential Energy tested almost 60,000 people in 23 countries to figure out what messages motivated them to act on climate change.[1] It found that the most effective motivator was not fear, anger, or even hope (which I've talked about a lot), but protecting what we love. "Protecting the planet for future generations" was the most unifying message. People love their kids and their grandkids. Even those who don't have children feel a responsibility to leave a good world for those who come after them. People feel a deep sense of connection with the environments they've grown up in: the seasons, the wildlife, the rivers, and the forests.

So, step one requires us to *want* to see change and be curious about what needs to be done. The fact that you've bought this book means you're probably already there. Step two is understanding the solutions we need and the challenges that come with each. I hope the 50 questions have given you clarity on that too. The next step, then, is to turn these insights into reality. I can't do that bit, but collectively we can. Whether we do so is up to us.

Further Reading

This book aims to provide sharp, concise answers to the wide range of questions I receive about how to solve climate change. It is deliberately broad rather than deep.

I could have written an entire book about any one of the sections—renewable energy, nuclear power, minerals, or geoengineering. Thankfully, many others have. Here is a short reading list of some of my favorite books on these topics, if you want to delve deeper.

Abundance—Ezra Klein and Derek Thompson. Despite being motivated to tackle climate change, the United States is moving extremely slowly on decarbonization. This is perhaps even more true in Democratic states, which seem incapable of building anything—grid expansion, solar farms, or high-speed rail—with any urgency. Ezra and Derek look at why this has happened and what it would take to turn things around.

Cobalt Red—Siddharth Kara. This book is a more detailed look at the human toll of mineral extraction. Siddharth looks at the impacts of cobalt mining—mostly for batteries in electronics and now electric vehicles—on workers, children, and communities in the Democratic Republic of Congo.

Electrify: An Optimist's Playbook for Our Clean Energy Future—Saul Griffith. Electrification is one the biggest levers we have to decarbonize the world (while saving money and energy at the same time). Saul provides a masterclass in why this is the case, and what an electrified world might look like.

Exponential: How Accelerating Technology Is Leaving Us Behind and What to Do About It—Azeem Azhar. To tackle climate change, we need to move quickly. Linear change simply won't be enough. That's why the rapid growth of technologies like solar, wind, and batteries makes me much more optimistic about our chances. Azeem Azhar explores the potential (and in some cases, risks) of exponential growth.

Geoengineering: The Gamble—Gernot Wagner. Geoengineering is easily the most controversial topic that I briefly covered in this book. Gernot tries to bring a balanced take on the very real risks and potential benefits of geoengineering in a warming world.

Acknowledgments

This book was born from the discussions I've had and burning questions I've answered over the last few years on how we tackle climate change. These questions have come from old and young, experts and non-experts, all united in their curiosity and determination to make a difference. So, to everyone who has engaged, challenged, and questioned: Thank you for the inspiration to put all the answers down on paper.

I feel extremely lucky to be supported by my team and colleagues at Our World in Data. I've learned so much from them and feel honored to work with people who are not only technically brilliant but also serious about tackling the world's biggest problems.

As always, my thanks go to my agent Toby Mundy and Grace Spencer at Aevitas Creative Management for having my back. To my editors at Chatto & Windus, Becky Hardie and Asia Choudhry: Thank you for sticking with me through the long journey of finding the right structure for my homeless words. To Katherine Fry for finding all of my poorly worded phrases, rogue commas, and mistakes. To Ruth Ellis for indexing. And a thank you to the rest of the team at Vintage, including Lucy Chaudhuri, Anna Redman Aylward, Lucy Beresford-Knox, and so many others working behind the scenes who don't get the credit they deserve. Thank you to my team at the MIT Press: Beth Clevenger, Katie Kerr, Judith Feldmann, and Natalie Jones. I appreciate you all.

There is an almost endless list of researchers, data nerds, and communicators on climate and energy that I've learned from. It's impossible to list them all, but here are just some of those that have helped shape my thinking on these issues: Kingsmill Bond, Jenny Chase, Zeke Hausfather, Michael Liebreich, Jan Rosenow, Auke Hoekstra, Nat Bullard, the team at Ember Energy, Chris Nelder, Bryony Worthington, David Ho, Helen Czerski, Jesse Jenkins, Mark Lynas, Erle Ellis, and Ken Caldeira. One of the reasons I feel optimistic about our prospects of transitioning to clean energy and solving climate change is that so many smart, generous people are working on it.

I've also learned a lot from Dave Reay, Satwat Rehman, and Scotland's Just Transition Commission. It's easy to get stuck on the data about electrons, carbon molecules, and solar panels; the work I do at the commission helps to remind me of the more complex pieces of the puzzle.

I can never pay enough thanks to my family and friends—particularly my parents—for the support, love, and deprecating humor (usually at my expense) when times get tough.

Finally, to my closest clan: Maisie and Cricket, who have perfect toe beans and noses for bopping, and Catherine, who is forever patient with my early morning typing and long silences when my head is stuck in a chapter. You're all very special to me.

I've dedicated this book to my niece, Maeva. Like many other toddlers, she will probably get to see the twenty-second century. What the world looks like then depends on the decisions that each of us makes today, tomorrow, and over the next few decades. It's a responsibility we should take seriously. I hope we don't let them down.

Notes

INTRODUCTION

1. P. Bradshaw, "Planet of the Humans Review—Contrarian Eco-Doc from the Michael Moore Stable," *The Guardian*, April 22, 2020, https://www.theguardian.com/film/2020/apr/22/planet-of-the-humans-review-environment-michael-moore-jeff-gibbs.
2. L. C. Stokes, "Michael Moore Produced a Film About Climate Change That's a Gift to Big Oil," *Vox*, April 8 2020, https://www.vox.com/2020/4/28/21238597/michael-moore-planet-of-the-humans-climate-change.
3. J. Simon, "In Some Fights over Solar, It's Environmentalist vs. Environmentalist," *NPR*, June 18, 2023, https://www.npr.org/2023/06/18/1177524841/solar-energy-project-location-debate; J. B. Ruhl and J. Salzman, "'The Greens' Dilemma: Building Tomorrow's Climate Infrastructure Today," *Emory Law Journal* 73 (2023).
4. Potential Energy Coalition, "The Road to Clean: How to Message Electric Vehicles in a Charged Environment," March 12, 2024, https://potentialenergycoalition.org/how-to-message-electric-vehicles-in-a-charged-environment; M. Smith, "What Role Should Nuclear Play in Britain's Climate Change Strategy?," *YouGov*, October 18, 2021, https://yougov.co.uk/politics/articles/38824-what-role-should-nuclear-play-britains-climate-cha; P. Andre et al., "Globally Representative Evidence on the Actual and Perceived Support for Climate Action," *Nature Climate Change* 14 (2024): 253–59.

QUESTION 1

1. International Energy Agency, *World Energy Outlook 2022* (2022); Climate Action Tracker, https://climateactiontracker.org/; UN Environment Programme, *Emissions Gap Report 2022: The Closing Window—Climate Crisis Calls for Rapid Transformation of Societies* (2022); S. Wang et al., "Mechanisms and Impacts of Earth System Tipping Elements," *Reviews of*

Geophysics 61, e2021RG000757 (2023); M. Meinshausen et al., "Realization of Paris Agreement Pledges May Limit Warming just Below 2 °C," *Nature* 604 (2022): 304–9; D.-J. van de Ven et al., "A Multimodel Analysis of Post-Glasgow Climate Targets and Feasibility Challenges," *Nature Climate Change* 13 (2023): 570–8.

2. Z. Hausfather and G. P. Peters, "Emissions—The 'Business as Usual' Story Is Misleading," *Nature* 577 (2020): 618–20; J. Ritchie and H. Dowlatabadi, "The 1000 GtC Coal Question: Are Cases of Vastly Expanded Future Coal Combustion Still Plausible?," *Energy Economics* 65 (2017): 16–31.
3. R. E. Kopp et al., "'Tipping Points' Confuse and Can Distract from Urgent Climate Action," *Nature Climate Change* (2024).
4. A. Jahn, M. M. Holland, and J. E. Kay, "Projections of an Ice-Free Arctic Ocean," *Nature Reviews Earth & Environment* 5, (2024): 164–76.
5. D. A. McKay, "Fact-Check: Will an Ice-Free Arctic Trigger a Climate Catastrophe?," April 2, 2019, https://climatetippingpoints.info/2019/04/02/fact-check-will-an-ice-free-arctic-trigger-a-climate-catastrophe/.

QUESTION 2

1. M. Vlasceanu et al., "Addressing Climate Change with Behavioral Science: A Global Intervention Tournament in 63 Countries," *Science Advances* 10 (2024): eadj5778.
2. A. Tyson and B. Kennedy, "How Americans View Future Harms from Climate Change in Their Community and Around the U.S.," Pew Research Center, October 25, 2023, https://www.pewresearch.org/science/2023/10/25/how-americans-view-future-harms-from-climate-change-in-their-community-and-around-the-u-s/.
3. European Commission, "Directorate General for Climate Action," *Climate Change: Report* (2023).
4. G. Sparkman, N. Geiger, and E. U. Weber, "Americans Experience a False Social Reality by Underestimating Popular Climate Policy Support by Nearly Half," *Nature Communications* 13 (2022): 4779.
5. Pew Research Center, "Public's Positive Economic Ratings Slip; Inflation Still Widely Viewed as Major Problem," May 23, 2024, https://www.pewresearch.org/politics/2024/05/23/top-problems-facing-the-u-s/.

QUESTION 3

1. US Energy Information Administration, Open Data, https://www.eia.gov/opendata/.
2. G. Dixon et al., "The Complexity of Pluralistic Ignorance in Republican Climate Change Policy Support in the United States," *Communications Earth & Environment* 5, 76 (2024).
3. A. Gustafson et al., "Republicans and Democrats Differ in Why They Support Renewable Energy," *Energy Policy* 141 (2020): 111448.

4. Potential Energy Coalition, *Later Is Too Late*, https://potentialenergycoalition.org/guides-and-reports/global-report/ (2023).

5. Potential Energy Coalition, "Can You Say 'Climate Change' to a Conservative?," https://potentialenergycoalition.org/can-you-say-climate-change-to-a-conservative/ (2023).

QUESTION 4

1. P. Friedlingstein et al., "Global Carbon Budget 2023," *Earth Systems Science Data* 15 (2023): 5301–69.

2. International Energy Agency, *Global EV Outlook 2024* (2024).

3. Ember, *Global Electricity Review 2024*, May 8, 2024, https://ember-climate.org/insights/research/global-electricity-review-2024/.

4. G. P. Peters, S. J. Davis, and R. Andrew, "A Synthesis of Carbon in International Trade," *Biogeosciences* 9 (2012): 3247–76.

QUESTION 5

1. National Minerals Information Center, "Mineral Commodity Summaries," US Geological Survey, https://www.usgs.gov/centers/national-minerals-information-center/mineral-commodity-summaries (2024).

2. Global Energy Monitor, *Boom and Bust Coal*, https://globalenergymonitor.org/report/boom-and-bust-coal-2024/ (2024).

3. Bloomberg News, "China's Wind and Solar Are Now Almost Enough to Power Every Home," November 13, 2023, https://www.bloomberg.com/news/articles/2023-02-14/china-s-wind-and-solar-are-now-almost-enough-to-power-every-home.

4. L. Bian et al., "China's Role in Accelerating the Global Energy Transition," Grantham Research Institute on Climate Change and the Environment, London School of Economics and Political Science (2024).

5. L. Myllyvirta, "Analysis: China's Emissions Set to Fall in 2024 After Record Growth in Clean Energy," *Carbon Brief*, November 13, 2023, https://www.carbonbrief.org/analysis-chinas-emissions-set-to-fall-in-2024-after-record-growth-in-clean-energy/.

PART II

1. V. Smil, *Harvesting the Biosphere: What We Have Taken from Nature* (MIT Press, 2013).

QUESTION 6

1. T. Moss and J. Kincer, "What Happens to Global Emissions if Africa Triples Down on Natural Gas for Power?," Energy for Growth Hub (2020).

2. V. Singh and K. Bond, "Powering Up the Global South: The Cleantech Path to Growth," *RMI*, October 2024, https://rmi.org/insight/powering-up-the-global-south/.

3. F. Guarascio, "EU Split Over Fertiliser Plants in Poorer Nations as Food Crisis Bites," Reuters, June 20, 2022, https://www.reuters.com/world/europe/eu-split-over-fertiliser-plants-poorer-nations-food-crisis-bites-2022-06-20/.

4. Food and Agriculture Organization of the United Nations, "Land, Inputs and Sustainability: Fertilizers by Nutrient" (2024).

5. Y. Osinbajo, "The Divestment Delusion: Why Banning Fossil Fuel Investments Would Crush Africa," *Foreign Affairs*, August 31, 2021, https://www.foreignaffairs.com/articles/africa/2021-08-31/divestment-delusion.

6. "Africa Being 'Punished' by Fossil Fuel Investment Ban—Niger," *Al Jazeera*, June 15, 2022, https://www.aljazeera.com/news/2022/6/15/africa-punished-by-investment-clamp-on-fossils-says-niger.

7. I. Mitchell and E. Wickstead, "Has the $100 Billion Climate Goal Been Reached?," Centre for Global Development (2024).

8. J. Gabbatiss, "Revealed: UK 'Double Counting' £500m of Aid for War-Torn Countries as Climate Finance," *Carbon Brief*, April 16, 2024, https://www.carbonbrief.org/revealed-uk-double-counting-500m-of-aid-for-war-torn-countries-as-climate-finance/.

9. D. Youcefi, "The Financial Cost of the Green Transition in Developing Countries," Finance for Development Lab (2022).

QUESTION 7

1. International Energy Agency, *Carbon Capture, Utilisation and Storage*, https://www.iea.org/energy-system/carbon-capture-utilisation-and-storage (2024).

2. A. Bacilieri, R. Black, and R. Way, "Assessing the Relative Costs of High-CCS and Low-CCS Pathways to 1.5 Degrees," Oxford Smith School, Working Paper 23-08 (2023).

3. International Energy Agency, *World Energy Investment 2023*, https://www.iea.org/reports/world-energy-investment-2023 (2023).

4. Z. Rempel, L. Cameron, and O. Bois von Kursk, "Unpacking Carbon Capture and Storage: The Technology Behind the Promise," International Institute for Sustainable Development (2023).

QUESTION 8

1. V. Smil, *Energy Transitions: History, Requirements, Prospects* (ABC-CLIO, 2010).

2. S. Evans, "'Exceptional New Normal': IEA Raises Growth Forecast for Wind and Solar by Another 25%," *Carbon Brief*, May 11, 2021, https://www.carbonbrief.org/exceptional-new-normal-iea-raises-growth-forecast-for-wind-and-solar-by-another-25/.

3. F. Creutzig et al., "The Underestimated Potential of Solar Energy to Mitigate Climate Change," *Nature Energy* 2 (2017): 17140; M. Jaxa-Rozen and E. Trutnevyte, "Sources of Uncertainty in Long-Term Global Scenarios of Solar Photovoltaic Technology," *Nature Climate Change* 11, 266–73 (2021).

QUESTION 9

1. Ember, *Half of the World Is Past a Peak in Fossil Power,* https://ember-energy.org/latest-insights/half-of-the-world-has-passed-peak-fossil-power/ (2023).
2. B. Kahn, "The World Hit 'Peak' Gas-Powered Vehicle Sales—in 2017," *Bloomberg,* January 30, 2024. https://www.bloomberg.com/news/articles/2024-01-30/world-hit-peak-gas-powered-vehicles-as-evs-gain-market-share.

QUESTION 10

1. US Department of Energy, "Where the Energy Goes: Electric Cars," https://www.fueleconomy.gov/feg/atv.shtml (accessed July 29, 2025).
2. N. Eyre, "From Using Heat to Using Work: Reconceptualising the Zero Carbon Energy Transition," *Energy Efficiency* 14 (2021): 77.

QUESTION 11

1. D. J. Murphy et al., "Energy Return on Investment of Major Energy Carriers: Review and Harmonization," *Sustainability* 14 (2022): 7098.
2. Fraunhofer Institute for Solar Energy Systems, ISE, *Photovoltaics Report* (2024).

QUESTION 12

1. K. Pahud and G. De Temmerman, "Overview of the EROI, a Tool to Measure Energy Availability Through the Energy Transition," *2022 8th International Youth Conference on Energy (IYCE)* (2022): 1–14.
2. A. de Vries, "The Growing Energy Footprint of Artificial Intelligence," *Joule* 7 (2023): 2191–94; International Energy Agency, *Electricity 2024* (2024); E. Masanet et al., "Recalibrating Global Data Center Energy-Use Estimates," *Science* 367 (2020): 984–86.
3. International Energy Agency, *World Energy Outlook 2024* (2024).
4. M. Leibreich, "Generative AI—The Power and the Glory," *BloombergNEF,* December 24, 2024. https://about.bnef.com/insights/clean-energy/liebreich-generative-ai-the-power-and-the-glory/.
5. T. Spencer and S. Singh, "What the Data Centre and AI Boom Could Mean for the Energy Sector," International Energy Agency, October 18, 2024. https://www.iea.org/commentaries/what-the-data-centre-and-ai-boom-could-mean-for-the-energy-sector.

6. C. Crownhart, "Why Microsoft Made a Deal to Help Restart Three Mile Island," *MIT Technology Review*, September 26, 2024.
7. E. Masanet et al., "Recalibrating Global Data Center Energy-Use Estimates," *Science* 367 (2020): 984–86.

QUESTION 13

1. International Energy Agency, *World Energy Employment 2023* (2023).
2. US Department of Energy, *United States Energy & Employment Report 2023*, https://www.energy.gov/policy/us-energy-employment-jobs-report-useer (2023).
3. PwC Green Jobs Barometer, "Green Jobs Growing at Four Times the Pace of the Overall Employment Market" (2022), https://www.pwc.co.uk/press-room/press-releases/green-jobs-growing-at-four-times-the-pace-of-the-overall-employm.html.
4. K. S. Gallagher and S. Oh, "Job Creation and Deep Decarbonization," *Oxford Review of Economic Policy* 39 (2023): 765–78.
5. C. Stewart, "Last UK Coal Plant Closes and Successfully Redeploys the Workforce," Trades Union Congress, September 26, 2024, https://www.tuc.org.uk/blogs/last-uk-coal-plant-closes-and-successfully-redeploys-workforce.

QUESTION 14

1. National Renewable Energy Laboratory, "Life Cycle Greenhouse Gas Emissions from Electricity Generation: Update" (2021).
2. UNECE, *Lifecycle Assessment of Electricity Generation Options* (2021).
3. S. Wang et al., "Future Demand for Electricity Generation Materials Under Different Climate Mitigation Scenarios," *Joule* (2023), https://doi.org/10.1016/j.joule.2023.01.001.

QUESTION 15

1. National Grid, "What Is Renewable Energy Storage (and Why Is It Important for Reaching Net Zero)?," https://www.nationalgrid.com/stories/energy-explained/what-is-renewable-energy-storage (2023).
2. National Grid Electricity System Operator (ESO), "Potential Electricity Storage Routes to 2050" (2020).
3. S. Farhad and A. Nazari, "Introducing the Energy Efficiency Map of Lithium-Ion Batteries," *International Journal of Energy Research* 43 (2019): 931–44.
4. A. Blakers et al., "A Review of Pumped Hydro Energy Storage," *Progress in Energy* 3 (2021): 022003.
5. UK Climate Change Committee, *The Sixth Carbon Budget: Electricity Generation* (2020).

6. Royal Society, *Large-Scale Electricity Storage*, royalsociety.org/electricity-storage (2023).
7. C. Goodall, *POSSIBLE: Ways to Net Zero* (Profile Books, 2024).

QUESTION 16

1. *The Economist*, "Sun, Wind and Drain," July 29, 2014, https://www.economist.com/finance-and-economics/2014/07/29/sun-wind-and-drain.
2. S. Evans, "Solar Is Now 'Cheapest Electricity in History,' Confirms IEA," *Carbon Brief*, October 13, 2024, https://www.carbonbrief.org/solar-is-now-cheapest-electricity-in-history-confirms-iea/.
3. R. Way et al., "Empirically Grounded Technology Forecasts and the Energy Transition," *Joule* 6 (2022): 2057–82.
4. International Energy Agency, *Renewables 2023* (2024).
5. S. Orangi et al., "Historical and Prospective Lithium-Ion Battery Cost Trajectories from a Bottom-Up Production Modeling Perspective," *Energy Storage* 76 (2024): 109800.
6. A. Colthorpe, "BloombergNEF: 'Already Cheaper to Install New-Build Battery Storage than Peaking Plants,'" *Energy Storage News* (2021); Lazard, *2023 Levelized Cost of Energy+*, https://www.lazard.com/research-insights/2023-levelized-cost-of-energyplus/ (2023).
7. Clean Energy Council, "Battery Storage: The New Clean Peaker," https://www.cleanenergycouncil.org.au/resources/resources-hub/battery-storage-the-new-clean-peaker (2021).
8. J. Lelieveld et al., "Effects of Fossil Fuel and Total Anthropogenic Emission Removal on Public Health and Climate," *Proceedings of the National Academy of Sciences* 116 (2019): 7192–97; K. Vohra et al., "Global Mortality from Outdoor Fine Particle Pollution Generated by Fossil Fuel Combustion: Results from GEOS-Chem," *Environmental Research* 195 (2021): 110754.
9. S. Black et al., "IMF Fossil Fuel Subsidies Data: 2023 Update" (2023).
10. G. Dolphin and Q. Xiahou, "World Carbon Pricing Database: Sources and Methods," *Scientific Data* 9 (2022): 573.
11. *BloombergNEF*, "2H 2022 Levelized Cost of Electricity Update" (2022).
12. F. J. M. M. Nijsse et al., "The Momentum of the Solar Energy Transition," *Nature Communications* 14 (2023): 6542.
13. F. C. Moore et al., "Synthesis of Evidence Yields High Social Cost of Carbon Due to Structural Model Variation and Uncertainties," *Proceedings of the National Academy of Sciences* 121 (2024): e2410733121.
14. E. Asdourian and D. Wessel, "What Is the Social Cost of Carbon?," Brookings, March 14, 2023, https://www.brookings.edu/articles/what-is-the-social-cost-of-carbon/.
15. T. Fisher, "The Political Economy of EPA's Updated Social Cost of Carbon," Cato Institute (2024).

16. W. J. Cole et al., "Quantifying the Challenge of Reaching a 100% Renewable Energy Power System for the United States," *Joule* 5 (2021): 1732–48.

QUESTION 17

1. H. Mirletz et al., "Unfounded Concerns About Photovoltaic Module Toxicity and Waste Are Slowing Decarbonization," *Nature Physics* 19 (2023): 1376–78.
2. M. Y. Khalid et al., "Recycling of Wind Turbine Blades Through Modern Recycling Technologies: A Road to Zero Waste," *Renewable Energy Focus* 44, 373–89 (2023).
3. B. Guezuraga, R. Zauner, and W. Pölz, "Life Cycle Assessment of Two Different 2 MW Class Wind Turbines," *Renewable Energy* 37 (2012): 37–44.

QUESTION 18

1. UNECE, *Lifecycle Assessment of Electricity Generation Options* (2021); J. Lovering et al., "Land-Use Intensity of Electricity Production and Tomorrow's Energy Landscape," *PLOS One* 17 (2022): e0270155.
2. Energy Transitions Commission, *Material and Resource Requirements for the Energy Transition*, https://www.energy-transitions.org/publications/material-and-resource-energy-transition (2023).
3. E. Larson et al., *Net-Zero America: Potential Pathways, Infrastructure, and Impacts, Final Report* (2021).
4. Joint Research Centre (European Commission), "Agrivoltaics Alone Could Surpass EU Photovoltaic 2030 Goals" (2023).

QUESTION 19

1. International Energy Agency, *Electricity Grids and Secure Energy Transitions*, https://www.iea.org/reports/electricity-grids-and-secure-energy-transitions (2023).
2. J. Rand et al., *Queued Up: Characteristics of Power Plants Seeking Transmission Interconnection* (2024).
3. J. P. Abbott, "Preparing for a Faster, More Efficient Electricity Connections Process," Ofgem, https://www.ofgem.gov.uk/blog/preparing-faster-more-efficient-electricity-connections-process (2024).
4. National Grid, "Queue Management: The Next Step in Accelerating Grid Connections," https://www.nationalgrid.com/electricity-transmission/queue-management-next-step-accelerating-grid-connections (2022).
5. International Energy Agency, *Electricity Grids and Secure Energy Transitions*, https://www.iea.org/reports/electricity-grids-and-secure-energy-transitions (2023).

6. *The Economist*, "Adding Capacity to the Electricity Grid Is Not a Simple Task," April 5, 2023, https://www.economist.com/technology-quarterly/2023/04/05/adding-capacity-to-the-electricity-grid-is-not-a-simple-task.

7. Department for Energy Security and Net Zero and Ofgem, *Connections Action Plan: Speeding Up Connections to the Electricity Network Across Great Britain* (2023).

8. S. Osaka, "How a Simple Fix Could Double the Size of the U.S. Electricity Grid," *Washington Post*, May 8, 2024, https://www.washingtonpost.com/climate-solutions/2024/05/28/reconductoring-us-electricity-grid-renewables/.

QUESTION 20

1. E. Stewart, "Donald Trump's Issue with Windmills Might Not Be About Birds," *Vox*, October 23, 2020, https://www.vox.com/policy-and-politics/2020/10/23/21530123/trump-presidential-debate-windmills-kill-birds.

2. L. Barrios and A. Rodríguez, "Behavioural and Environmental Correlates of Soaring-Bird Mortality at On-Shore Wind Turbines," *Journal of Applied Ecology* 41 (2004): 72–81; C. B. Thaxter et al., "Bird and Bat Species' Global Vulnerability to Collision Mortality at Wind Farms Revealed Through a Trait-Based Assessment," *Proceedings of the Royal Society B: Biological Sciences* 284 (2017): 20170829.

3. R. M. R. Barclay, E. F. Baerwald, and J. C. Gruver, "Variation in Bat and Bird Fatalities at Wind Energy Facilities: Assessing the Effects of Rotor Size and Tower Height," *Canadian Journal of Zoology* 85 (2007): 381–87.

4. E. B. Arnett et al., "Altering Turbine Speed Reduces Bat Mortality at Wind-Energy Facilities," *Frontiers in Ecology and the Environment* 9 (2011): 209–14.

5. E. M. Bennett et al., "Curtailment as a Successful Method for Reducing Bat Mortality at a Southern Australian Wind Farm," *Austral Ecology* 47 (2022): 1329–39.

6. M. Subramanian, "The Trouble with Turbines: An Ill Wind," *Nature* 486 (2012): 310–11.

7. R. May et al., "Paint It Black: Efficacy of Increased Wind Turbine Rotor Blade Visibility to Reduce Avian Fatalities," *Ecology and Evolution* 10 (2020): 8927–35.

8. S. P. Weaver et al., "Ultrasonic Acoustic Deterrents Significantly Reduce Bat Fatalities at Wind Turbines," *Global Ecology and Conservation* 24 (2020): e01099.

9. E. Bennett, "Overkill or Industrial Best Practise? The Impact of Wind Farms on Birds and Bats." https://assets.cleanenergycouncil.org.au/documents/events/event-docs-2019/WIF19/2-Emma-Bennett.pdf (2019); J. Merriman, "How Many Birds Are Killed by Wind Turbines?," *American Bird Conservancy* (2021); S. R. Loss, T. Will, and P. P. Marra, "Direct Mortality of Birds from Anthropogenic Causes," *Annual Review of Ecology, Evolution, and Systematics* 46 (2015): 99–120.

QUESTION 21

1. Our World in Data, "What Are the Safest and Cleanest Sources of Energy?," https://ourworldindata.org/safest-sources-of-energy (2020).
2. A. Markandya and P. Wilkinson, "Electricity Generation and Health," *Lancet* 370 (2007): 979–90.
3. H. Ritchie, "What Was the Death Toll from Chernobyl and Fukushima?," Our World in Data, https://ourworldindata.org/what-was-the-death-toll-from-chernobyl-and-fukushima (2017).
4. M. Hayakawa, "Increase in Disaster-Related Deaths: Risks and Social Impacts of Evacuation," *Annals of the ICRP* 45 (2016): 123–8.
5. UNSCEAR, "Report of the United Nations Scientific Committee on the Effects of Atomic Radiation" (2013).

QUESTION 22

1. International Atomic Energy Agency, "The Database on Nuclear Power Reactors," Power Reactor Information System (PRIS), https://pris.iaea.org/pris/Home.aspx (accessed 2023).
2. International Atomic Energy Agency, *Nuclear Power Reactors in the World: 2021 Edition* (2021).
3. International Energy Agency, *Nuclear Power in a Clean Energy System*, https://www.iea.org/reports/nuclear-power-in-a-clean-energy-system (2019).
4. M. Moore, "Google Orders Small Modular Nuclear Reactors for Its Data Centres," *Financial Times*, October 14, 2024, https://www.ft.com/content/29eaf03f-4970-40da-ae7c-c8b3283069da.
5. B. Flyvbjerg and D. Gardner, *How Big Things Get Done: The Surprising Factors That Determine the Fate of Every Project, from Home Renovations to Space Exploration and Everything in Between* (Currency, 2023).

QUESTION 23

1. J. R. Lovering, A. Yip, and T. Nordhaus, "Historical Construction Costs of Global Nuclear Power Reactors', *Energy Policy* 91 (2016): 371–82.
2. World Nuclear News, "AP1000 Remains Attractive Option for US Market, Says MIT," April 8, 2022, https://world-nuclear-news.org/Articles/AP1000-remains-attractive-option-for-US-market-say.
3. P. Eash-Gates et al., "Sources of Cost Overrun in Nuclear Power Plant Construction Call for a New Approach to Engineering Design," *Joule* 4 (2020): 2348–73.
4. Nuclear Energy Agency, *Unlocking Reductions in the Construction Costs of Nuclear: A Practical Guide for Stakeholders* (2020).

5. B. Potter, "Why Does Nuclear Power Plant Construction Cost So Much?," *Institute for Progress*, May 1, 2023, https://ifp.org/nuclear-power-plant-construction-costs/.

6. US Department of Energy, *Pathways to Commercial Liftoff: Advanced Nuclear* (2023).

7. S. Dumitriu and B. Hopkinson, "Infrastructure Costs: Nuclear Edition," *Notes on Growth*, December 4, 2023, https://www.samdumitriu.com/p/infrastructure-costs-nuclear-edition.

8. International Energy Agency, *Projected Costs of Generating Electricity 2020*, https://www.iea.org/reports/projected-costs-of-generating-electricity-2020 (2020).

QUESTION 24

1. US Department of Energy, "7 Things The Simpsons Got Wrong About Nuclear," https://www.energy.gov/ne/articles/7-things-simpsons-got-wrong-about-nuclear (2018).

2. International Atomic Energy Agency, *Status and Trends in Spent Fuel and Radioactive Waste Management* (2022).

3. S. Krikorian, "France's Efficiency in the Nuclear Fuel Cycle: What Can 'Oui' Learn?," International Atomic Energy Agency, September 4, 2019, https://www.iaea.org/newscenter/news/frances-efficiency-in-the-nuclear-fuel-cycle-what-can-oui-learn.

4. C. Clifford, "The Energy in Nuclear Waste Could Power the U.S. for 100 Years, but the Technology Was Never Commercialized," *CNBC*, June 2, 2022, https://www.cnbc.com/2022/06/02/nuclear-waste-us-could-power-the-us-for-100-years.html.

QUESTION 25

1. G. Bieker, *A Global Comparison of the Life-Cycle Greenhouse Gas Emissions of Combustion Engine and Electric Passenger Cars*, ICCT (2021).

2. Ricardo, *Lifecycle Analysis of UK Road Vehicles* (2021).

3. Benchmark Source, "How Low Can Battery Emissions Go?," May 2, 2024, https://source.benchmarkminerals.com/article/how-low-can-battery-emissions-go.

4. S. Evans, "Factcheck: 21 Misleading Myths About Electric Vehicles," *Carbon Brief*, October, 24, 2023, https://www.carbonbrief.org/factcheck-21-misleading-myths-about-electric-vehicles/.

5. E. Evrard et al., *Carbon Footprint Report—Volvo C40 Recharge* (2021).

6. US Environment Protection Agency, "Electric Vehicle Myths," https://www.epa.gov/greenvehicles/electric-vehicle-myths (accessed 2024).

7. Zeke Hausfather, "Factcheck: How Electric Vehicles Help to Tackle Climate Change," *Carbon Brief*, May 13, 2019, https://www.carbonbrief.org/factcheck-how-electric-vehicles-help-to-tackle-climate-change/; International Transport Forum, *Life-Cycle Assessment of Passenger Transport: An Indian Case Study* (2023).

8. J. Ewing, "Volvo Plans to Sell Only Electric Cars by 2030," *New York Times*, October 22, 2021, https://www.nytimes.com/2021/03/02/business/volvo-electric-cars.html.

QUESTION 26

1. International Transport Forum, *Life-Cycle Assessment of Passenger Transport: An Indian Case Study* (2023); D. R. Peters et al., "Public Health and Climate Benefits and Trade-Offs of U.S. Vehicle Electrification," *GeoHealth* (2020); A. Sen, "Why Will Electric Vehicles Benefit India Irrespective of New Power Sector Policies?," International Council on Clean Transportation (2021); Berkeley Lab, *2035 The Report. Plummeting Costs and Dramatic Improvements in Batteries Can Accelerate Our Clean Transportation Future* (2021); W.-Y. Lin et al., "Analysis of Air Quality and Health Co-Benefits Regarding Electric Vehicle Promotion Coupled with Power Plant Emissions," *Journal of Cleaner Production* 247 (2020):119152.
2. T. Grigoratos et al., "Interlaboratory Study on Brake Particle Emissions—Part I: Particulate Matter Mass Emissions," *Atmosphere* 14 (2023): 498.

QUESTION 27

1. Rutgers, "The Biggest Barrier to a Vibrant Second-Hand EV Market? Price," *ScienceDaily*, April 19, 2024, https://www.sciencedaily.com/releases/2024/04/240419132743.htm; S&P Global Mobility, "Affordability Tops Charging and Range Concerns in Slowing EV Demand" (2023); *Washington Post*, "July 13–23, 2023, Washington Post-University of Maryland Climate Poll" (2023); DIA, "Over Half of Drivers Reveal They Won't Buy an Electric Car—and Cost Is the Main Barrier" (2023).
2. International Energy Agency, "Electric Vehicles: Total Cost of Ownership Tool," https://www.iea.org/data-and-statistics/data-tools/electric-vehicles-total-cost-of-ownership-tool (2022); C. Harto, *Electric Vehicle Ownership Costs: Today's Electric Vehicles Offer Big Savings for Consumers* (2020); P. Campbell, "Electric Car Costs Draw Level with Petrol and Diesel," *Financial Times*, December 11, 2022, https://www.ft.com/content/24ae34f9-2d61-45cb-9bbc-158c2e5e26b3; F. Sperka, "Electric Cars Are Still Cheaper to Run than Petrol and Diesel. T&E Did the Maths," Transport and Environment (2022).
3. G. Smeeton, "New Analysis: Petrol Car Drivers Paid a £700 'Petrol Premium' in 2023," Energy & Climate Intelligence Unit (2023).
4. Gulf Oil, "The Cost of Running an Electric Vehicle vs Petrol Car in 33 Countries," September 9, 2023, https://www.gulfoilltd.com/exploring-ev-and-petrol-running-costs-across-nations.
5. F. Lambert, "Tesla Has Lowest Maintenance and Repair Cost of Any Brand," *Electrek*, April 22, 2024, https://electrek.co/2024/04/22/tesla-lowest-maintenance-repair-cost-any-brand/.
6. *BloombergNEF*, "Electric Cars to Reach Price Parity by 2025," July 23, 2017, https://about.bnef.com/blog/electric-cars-reach-price-parity-2025/.
7. N. Rivero, "It's Never Been Cheaper to Buy an EV. Here's Why," *Washington Post*, May 18, 2024, https://www.washingtonpost.com/climate-solutions/2024/03/18/electric-vehicle-price-drop/.

8. H. Ritchie, "Many Second-Hand Electric Cars Are Cheaper Up-Front Than Their Petrol Equivalents," *Sustainability by Numbers*, April 2, 2025, https://www.sustainabilitybynumbers.com/p/used-electric-car-costs.

QUESTION 28

1. International Energy Agency, "Evolution of Average Range of Electric Vehicles by Powertrain, 2010–2021" (2022).
2. Fully Charged, Electric Vehicle Database, https://ev-database.org/uk (2023).

QUESTION 29

1. L. Fischer et al., "Exploring Consumer Sentiment on Electric-Vehicle Charging," McKinsey & Company (2024).
2. W. Sierzchula et al., "The Influence of Financial Incentives and Other Socio-Economic Factors on Electric Vehicle Adoption," *Energy Policy* 68 (2014): 183–94; D. Hall and N. Lutsey, *Emerging Best Practices for Electric Vehicle Charging Infrastructure* (2017).
3. Transport & Environment, "Public Charging in Europe: Where Are We At?," April 17, 2024, https://www.transportenvironment.org/articles/public-charging-in-europe-where-are-we-at.
4. S. LaMonaca and L. Ryan, "The State of Play in Electric Vehicle Charging Services—A Review of Infrastructure Provision, Players, and Policies," *Renewable and Sustainable Energy Reviews* 154 (2022): 111733.
5. J. G. Smart and S. D. Salisbury, *Plugged In: How Americans Charge Their Electric Vehicles*, https://doi.org/10.2172/1369632 (2015).
6. L. Fischer et al., "Exploring Consumer Sentiment on Electric-Vehicle Charging," McKinsey & Company (2024).

QUESTION 30

1. H. Ritchie, "How Much More Electricity Will the UK Need to Switch to Electric Vehicles?," *Sustainability by Numbers*, September 15, 2023, https://www.sustainabilitybynumbers.com/p/uk-ev-electricity-demand.
2. J. Morris, "Electricity Grids Can Handle Electric Vehicles Easily—They Just Need Proper Management," *Forbes*, November 13, 2021, https://www.forbes.com/sites/jamesmorris/2021/11/13/electricity-grids-can-handle-electric-vehicles-easily--they-just-need-proper-management/.
3. C. O'Malley et al., "Frequency Response from Aggregated V2G Chargers With Uncertain EV Connections," *IEEE Transactions on Power Systems*, 1–14 (2022), https://doi.org/10.1109/TPWRS.2022.3202607.

4. N. O. Nagel, E. O. Jåstad, and T. Martinsen, "The Grid Benefits of Vehicle-to-Grid in Norway and Denmark: An Analysis of Home- and Public Parking Potentials," *Energy* 293, 130729 (2024).

QUESTION 31

1. Norwegian Automobile Association (NAF), "Electric Car Test," https://www.naf.no/elbil/elprix (2022).

QUESTION 32

1. M. Sundin, "Electric Cars Catch Fire Less Often than Fossil Fuel Cars," *Warp News*, May 10, 2023, https://www.warpnews.org/transportation/electric-cars-catch-fire-less-often-than-fossil-fuel-cars/.
2. "EV FireSafe: Enhancing Safety for Emergency Responders at Electric Vehicle Traction Battery Fires," https://www.evfiresafe.com/ (accessed 2024).

QUESTION 33

1. International Energy Agency, *The Role of Critical Minerals in Clean Energy Transitions*, https://www.iea.org/reports/the-role-of-critical-minerals-in-clean-energy-transitions (2021); K. Mackenzie, "World Lacks Time, Not Minerals for Climate-Saving Technology," Bloomberg (2022); K. Bond and S. Butler-Sloss, *The Renewable Revolution: It's Exponential, Global, and This Decade* (RMI, 2023).
2. S. Wang et al., "Future Demand for Electricity Generation Materials Under Different Climate Mitigation Scenarios," *Joule* (2023), https://doi.org/10.1016/j.joule.2023.01.001.
3. Colorado School of Mines, *The State of Critical Minerals Report 2023*, https://payneinstitute.mines.edu/the-state-of-critical-minerals-report-2023/ (2023).

QUESTION 34

1. S. Wang et al., "Updated Mining Footprints and Raw Material Needs for Clean Energy," Breakthrough Institute, April 25, 2024, https://thebreakthrough.org/issues/energy/updated-mining-footprints-and-raw-material-needs-for-clean-energy.
2. International Energy Agency, "Mineral Requirements for Clean Energy Transitions—The Role of Critical Minerals in Clean Energy Transitions," https://www.iea.org/reports/the-role-of-critical-minerals-in-clean-energy-transitions/mineral-requirements-for-clean-energy-transitions (2022).
3. J. Nijnens et al., "Energy Transition Will Require Substantially Less Mining than the Current Fossil System," *Joule* 7, 2408–13 (2023).
4. T. Watari et al., "Sustainable Energy Transitions Require Enhanced Resource Governance," *Journal of Cleaner Production* 312 (2021): 127698.

QUESTION 35

1. USGS, Cobalt Statistics and Information, https://www.usgs.gov/centers/national-minerals-information-center/cobalt-statistics-and-information (accessed 2023).

2. S. Kara, *Cobalt Red: How the Blood of the Congo Powers Our Lives* (St. Martin's Press, 2023).

3. BloombergNEF, "Race to Net Zero: The Pressures of the Battery Boom in Five Charts," July 15, 2022, https://www.bloomberg.com/professional/insights/commodities/race-to-net-zero-pressures-of-the-battery-boom-in-five-charts/.

4. M. D. Bazilian, "The Mineral Foundation of the Energy Transition," *Extractive Industries and Society* 5 (2018): 93–97.

5. Natural Resource Governance Institute, "Preventing Corruption in Energy Transition Mineral Supply Chains," https://resourcegovernance.org/publications/preventing-corruption-energy-transition-mineral-supply-chains (2022).

6. Cobalt Institute, "Responsible Sourcing," https://www.cobaltinstitute.org/cobalt-sourcing-responsibility/ (2023).

QUESTION 36

1. International Energy Agency, "Mineral Requirements for Clean Energy Transitions—The Role of Critical Minerals in Clean Energy Transitions," https://www.iea.org/reports/the-role-of-critical-minerals-in-clean-energy-transitions/mineral-requirements-for-clean-energy-transitions (2022).

2. Y. Guan et al., "Burden of the Global Energy Price Crisis on Households," *Nature Energy* 8 (2023): 304–16.

3. *The Economist*, "Expensive Energy May Have Killed More Europeans Than Covid-19 Last Winter," May 10, 2023, https://www.economist.com/graphic-detail/2023/05/10/expensive-energy-may-have-killed-more-europeans-than-covid-19-last-winter.

4. J. Krane and R. Idel, "More Transitions, Less Risk: How Renewable Energy Reduces Risks from Mining, Trade and Political Dependence," *Energy Research and Social Science* 82 (2021): 102311.

QUESTION 37

1. L. Wang et al., "Life Cycle Water Use of Gasoline and Electric Light-Duty Vehicles in China," *Resources Conservation and Recycling* 154 (2020): 104628.

2. M. Noori, S. Gardner, and O. Tatari, "Electric Vehicle Cost, Emissions, and Water Footprint in the United States: Development of a Regional Optimization Model," *Energy* 89 (2015): 610–25; N. C. Onat, M. Kucukvar, and O. Tatari, "Well-to-Wheel Water Footprints of Conventional Versus Electric Vehicles in the United States: A State-Based Comparative Analysis," *Journal of Cleaner Production* 204 (2018): 788–802; B. Holmatov and A. Y. Hoekstra, "The

Environmental Footprint of Transport by Car Using Renewable Energy," *Earth's Future* 8 (2020): e2019EF001428.

3. Y. Jin et al., "Water Use of Electricity Technologies: A Global Meta-Analysis," *Renewable and Sustainable Energy Reviews* 115 (2019): 109391.
4. US Energy Information Administration, "Water Withdrawals by U.S. Power Plants Have Been Declining," *Today in Energy*, November 9, 2018, https://www.eia.gov/todayinenergy/detail.php?id=37453; T. de O. Bredariol, "Clean Energy Can Help to Ease the Water Crisis," International Energy Agency (2023).
5. S. Lakshman, "More Critical Minerals Mining Could Strain Water Supplies in Stressed Regions," World Resources Institute (2024).
6. N. Ramirez, "Modernization of Chile's Water Code," USDA (2022); US International Trade Administration, "Chile—Country Commercial Guide," https://www.trade.gov/country-commercial-guides/chile-mining (2023).

PART VII

1. International Energy Agency, "Heating," https://www.iea.org/energy-system/buildings/heating (2023).
2. International Energy Agency, *Energy Technology Perspectives 2023*, https://www.iea.org/reports/energy-technology-perspectives-2023 (2023).

QUESTION 38

1. European Heat Pump Association, Market data, https://www.ehpa.org/market-data/ (2024).
2. D. Gibb et al., "Coming in from the Cold: Heat Pump Efficiency at Low Temperatures," *Joule* 7 (2023): 1939–42.
3. Nesta, "Heat Pumps: A User Survey," May 23, 2023, https://www.nesta.org.uk/report/heat-pumps-a-user-survey/.
4. Nesta, "How to Scale a Highly Skilled Heat Pump Industry," July 7, 2022, https://www.nesta.org.uk/report/how-to-scale-a-highly-skilled-heat-pump-industry/.

QUESTION 39

1. UK Department for Energy Security and Net Zero, Press release: "Heat Pump Grants Increased by 50%" (2023).
2. International Energy Agency, "The Future of Heat Pumps," https://www.iea.org/reports/the-future-of-heat-pumps (2022).
3. L. Orso and M. Gabriel, "The Electricity-to-Gas Price Ratio Explained—How a 'Green Ratio' Would Make Bills Cheaper and Greener," Nesta (2023).

4. E. J. Wilson, P. Munankarmi, B. D. Less, J. L. Reyna, and S. Rothgeb, "Heat Pumps for All? Distributions of the Costs and Benefits of Residential Air-Source Heat Pumps in the United States," *Joule* 8, no. 4 (2024): 1000–35.
5. M. Aunedi et al., "System-Driven Design and Integration of Low-Carbon Domestic Heating Technologies," *Renewable and Sustainable Energy Reviews* 187 (2023): 113695; National Infrastructure Commission, "Technical Annex: Hydrogen Heating," https://nic.org.uk/app/uploads/NIA-2-Technical-annex-hydrogen-heating-Final-18-October-2023.pdf (2023).
6. P. Heptonstall and M. Winskel, "Decarbonising Home Heating: An Evidence Review of Domestic Heat Pump Installed Costs," https://ukerc.ac.uk/publications/heat-pump-cost-review/ (2023).

QUESTION 40

1. R. Mutino et al., "Air-Conditioning Should Be a Human Right in the Climate Crisis," *Scientific American*, May 10, 2022, https://www.scientificamerican.com/article/air-conditioning-should-be-a-human-right-in-the-climate-crisis/.
2. International Energy Agency, *The Future of Cooling*, https://www.iea.org/reports/the-future-of-cooling (2018).
3. L. Davis et al., "Air conditioning and global inequality," *Global Environmental Change* 69 (2021): 102299.
4. N. Howarth et al., "Keeping Cool in a Hotter World Is Using More Energy, Making Efficiency More Important than Ever," International Energy Agency (2023).
5. J. Woods et al., "Humidity's Impact on Greenhouse Gas Emissions from Air Conditioning," *Joule* 6 (2022): 726–41.
6. R. Opoku, I. A. Edwin, and K. A. Agyarko, "Energy Efficiency and Cost Saving Opportunities in Public and Commercial Buildings in Developing Countries—The Case of Air-Conditioners in Ghana," *Journal of Cleaner Production* 230, 937–44 (2019).
7. UN Environment Programme & International Energy Agency, *Cooling Emissions and Policy Synthesis Report* (2020).

PART VIII

1. J. Poore & T. Nemecek, "Reducing Food's Environmental Impacts Through Producers and Consumers," *Science* 360 (2018): 987–92; M. Crippa et al., "Food Systems Are Responsible for a Third of Global Anthropogenic GHG Emissions," *Nature Food* 2 (2021): 198–209.

QUESTION 41

1. FAOSTAT, https://www.fao.org/faostat/en/#data (2024).
2. W. Fraanje and T. Garnett, "Soy: Food, Feed, and Land Use Change," Food Climate Research Network, University of Oxford (2020).

3. A. Shepon et al., "Energy and Protein Feed-to-Food Conversion Efficiencies in the US and Potential Food Security Gains from Dietary Changes," *Environmental Research Letters* 11, 105002 (2016).

4. J. Poore & T. Nemecek, "Reducing Food's Environmental Impacts Through Producers and Consumers," *Science* 360 (2018): 987–92.

QUESTION 42

1. P. Sinke et al., "Ex-Ante Life Cycle Assessment of Commercial-Scale Cultivated Meat Production in 2030," *International Journal of Life Cycle Assessment* 28 (2023): 234–54; H. L. Tuomisto and M. J. Teixeira De Mattos, "Environmental Impacts of Cultured Meat Production," *Environmental Science and Technology* 45 (2011): 6117–23.

2. A. Klein, "Lab-Grown Meat Could Be 25 Times Worse for the Climate than Beef," *New Scientist*, May 9, 2023, https://www.newscientist.com/article/2372229-lab-grown-meat-could-be-25-times-worse-for-the-climate-than-beef/; D. Risner et al., "Environmental Impacts of Cultured Meat: A Cradle-to-Gate Life Cycle Assessment," preprint at https://doi.org/10.1101/2023.04.21.537778 (2023).

QUESTION 43

1. A. J. Kortleve et al., "Over 80% of the European Union's Common Agricultural Policy Supports Emissions-Intensive Animal Products," *Nature Food* 5 (2024): 288–92.

2. YouGov, YouGov/Lexington Communications Survey Results, https://docs.cdn.yougov.com/z1mtmetxbm/LexComm_PlantbasedFoods_210812_W.pdf (2021).

3. H. Dimbleby, *National Food Strategy: Independent Review*, https://www.nationalfoodstrategy.org/ (2021).

4. M. Springmann and F. Freund, "Options for Reforming Agricultural Subsidies from Health, Climate, and Economic Perspectives," *Nature Communications* 13, 82 (2022).

5. Good Food Institute, "The science of fermentation," https://gfi.org/science/the-science-of-fermentation/ (2023).

QUESTION 44

1. Y. Wang and C. Jian, "Novel Plant-Based Meat Alternatives: Implications and Opportunities for Consumer Nutrition and Health," in *Advances in Food and Nutrition Research*, vol. 106 (Elsevier, 2023), 241–74.

2. A. A. Coffey, R. Lillywhite, and O. Oyebode, "Meat Versus Meat Alternatives: Which Is Better for the Environment and Health? A Nutritional and Environmental Analysis of Animal-Based Products Compared with Their Plant-Based Alternatives," *Journal of Human Nutrition and Dietetics* 36 (2023): 2147–56.

3. A. K. Roberts et al., "SWAP-MEAT Athlete (study with appetizing plant-food, meat eating alternatives trial)—investigating the impact of three different diets on recreational athletic

performance: a randomized crossover trial," *Nutrition Journal* 21 (2022): 69; V. Hevia-Larraín et al., "High-Protein Plant-Based Diet Versus a Protein-Matched Omnivorous Diet to Support Resistance Training Adaptations: A Comparison Between Habitual Vegans and Omnivores," *Sports Medicine* 51 (2021): 1317–30.

4. Z. Fang et al., "Association of Ultra-Processed Food Consumption with All Cause and Cause Specific Mortality: Population Based Cohort Study," *BMJ* e078476, https://doi.org/10.1136/bmj-2023-078476 (2024).

5. R. Cordova et al., "Consumption of Ultra-Processed Foods and Risk of Multimorbidity of Cancer and Cardiometabolic Diseases: A Multinational Cohort Study," *Lancet Regional Health—Europe* 35 (2023): 100771.

6. L. Sanchez-Siles et al., "Naturalness and Healthiness in 'Ultra-Processed Foods': A Multidisciplinary Perspective and Case Study," *Trends in Food Science and Technology* 129 (2022): 667–73.

7. M. Song et al., "Association of Animal and Plant Protein Intake with All-Cause and Cause-Specific Mortality," *JAMA Internal Medicine* 176 (2016): 1453; Y. Sun et al., "Association of Major Dietary Protein Sources with All-Cause and Cause-Specific Mortality: Prospective Cohort Study," *Journal of the American Heart Association* 10 (2021): e015553.

QUESTION 45

1. UN-Habitat, *World Cities Report 2022: Envisaging the Future of Cities* (2022).

2. International Energy Agency, "Cement Technology Roadmap Plots Path to Cutting CO2 Emissions 24% by 2050" (2018).

3. Alliance for Low-Carbon Cement and Concrete, "Fast-Tracking Cement Decarbonisation: From Underperforming to Performance-Based Standards" (2023).

4. B. Gates, *How to Avoid a Climate Disaster: The Solutions We Have and the Breakthroughs We Need* (Allen Lane, 2021).

5. Leilac, "Leilac-2 passes Financial Investment Decision" (2022).

QUESTION 46

1. International Energy Agency, *Iron and Steel Technology Roadmap*, https://www.iea.org/reports/iron-and-steel-technology-roadmap (2020).

2. US Energy Information Administration, "EIA Analysis Explores Energy Effects of Early Adoption of Low-Carbon Steelmaking" (2022).

3. C. Goodall, *POSSIBLE: Ways to Net Zero* (Profile Books, 2024).

4. Reuters, "Sweden's H2 Green Steel Raises $5.2 bln in New Funding," January 22, 2024, https://www.reuters.com/business/swedens-h2-green-steel-raises-52-bln-new-funding-2024-01-22/.

5. Energy Transitions Commission, "Steel," https://www.energy-transitions.org/sector/industry/steel/ (2023).
6. Energy Transitions Commission, *Net-Zero Steel Sector Sector Transition Strategy* (2021).

QUESTION 47

1. D. S. Lee et al., "The Contribution of Global Aviation to Anthropogenic Climate Forcing for 2000 to 2018," *Atmospheric Environment* 244 (2021): 117834.
2. IATA, "Our Commitment to Fly Net Zero by 2050," https://www.iata.org/en/programs/environment/flynetzero/ (accessed 2024).
3. UN Trade and Development, *Review of Maritime Transport 2023*, https://unctad.org/publication/review-maritime-transport-2023 (2023).
4. M. Liebreich, "Liebreich: Net Zero Will Be Harder Than You Think—And Easier. Part II: Easier," *BloombergNEF*, February 22, 2024, https://about.bnef.com/insights/clean-energy/liebreich-net-zero-will-be-harder-than-you-think-and-easier-part-ii-easier/; M. Barnard, "How Will Climate Action Change the Face of Global Shipping?," *Forbes*, December 5, 2023, https://www.forbes.com/sites/michaelbarnard/2023/12/05/how-will-climate-action-change-the-face-of-global-shipping/.
5. S. R. Sæther & E. Moe, "A Green Maritime Shift: Lessons from the Electrification of Ferries in Norway," *Energy Research and Social Science* 81 (2021): 102282.
6. Mærsk Mc-Kinney Møller Center for Zero Carbon Shipping & Global Maritime Forum, "The Shipping Industry's Fuel Choices on the Path to Net Zero," https://www.globalmaritimeforum.org/content/2023/04/the-shipping-industrys-fuel-choices-on-the-path-to-net-zero_final.pdf (2023).
7. Lloyd's Register, *Fuel for Thought: Methanol for Passenger Ships* (2024).
8. IRENA and Methanol Institute, *Innovation Outlook: Renewable Methanol*, International Renewable Energy Agency (2021).
9. C. Bergero et al., "Pathways to Net-Zero Emissions from Aviation," *Nature Sustainability* 6 (2023): 404–14.
10. Bergero et al., "Pathways."
11. Maersk, "Maersk Names First Vessel of Its Large Methanol-Enabled Fleet 'Ane Maersk,'" https://www.maersk.com/news/articles/2024/01/26/maersk-names-first-vessel-of-its-large-methanol-enabled-fleet-ane-maersk (2024).

QUESTION 48

1. R. Prütz et al., "Understanding the Carbon Dioxide Removal Range in 1.5°C Compatible and High Overshoot Pathways," *Environmental Research Communications* 5 (2023): 041005; H. J. Buck et al., "Why Residual Emissions Matter Right Now," *Nature Climate Change* 13 (2023): 351–58.

2. D. T. Ho, "Carbon Dioxide Removal Is Not a Current Climate Solution—We Need to Change the Narrative," *Nature* 616 (2023): 9.

3. T. Chen, "2023 Investment Landscape in Carbon Removal," *CDR.fyi*, January 11, 2024, https://www.cdr.fyi/blog/2023-investment-landscape-in-carbon-removal.

QUESTION 49

1. J. Gertner, "The Tiny Swiss Company That Thinks It Can Help Stop Climate Change," *New York Times*, February 12, 2019, https://www.nytimes.com/2019/02/12/magazine/climeworks-business-climate-change.html; L. Hook, "World's Biggest 'Direct Air Capture' Plant Starts Pulling in CO_2," *Financial Times*, September 8, 2021, https://www.ft.com/content/8a942e30-0428-4567-8a6c-dc704ba3460a.

2. T. Schreiter, "Will Direct Air Capture Ever Be Affordable? The Rise of DAC 3.0," Extantia Capital, https://medium.com/extantia-capital/will-direct-air-capture-ever-be-affordable-the-rise-of-dac-3-0-d46cb355f35c (2024).

3. J. Young et al., "The Cost of Direct Air Capture and Storage Can Be Reduced via Strategic Deployment but Is Unlikely to Fall Below Stated Cost Targets," *One Earth* 6 (2023): 899–917; K. Sievert, T. S. Schmidt, and B. Steffen, "Considering Technology Characteristics to Project Future Costs of Direct Air Capture," *Joule* 8 (2024): 979–99.

QUESTION 50

1. W. Smith and G. Wagner, "Stratospheric Aerosol Injection Tactics and Costs in the First 15 Years of Deployment," *Environmental Research Letters* 13, 124001 (2018).

2. L. Burns et al., "Technology Factsheet Series: Solar Geoengineering," Belfer Center for Science and International Affairs, Harvard Kennedy School (2019).

3. W. Smith, "The Cost of Stratospheric Aerosol Injection Through 2100," *Environmental Research Letters* 15 (2020): 114004.

4. A. Parker and P. J. Irvine, "The Risk of Termination Shock from Solar Geoengineering," *Earth's Future* 6 (2018): 456–67.

5. E. Conway, "Just 5 Questions: Hacking the Planet," NASA Jet Propulsion Laboratory (2014), https://climate.nasa.gov/news/1066/just-5-questions-hacking-the-planet/.

6. A. A. Rahman et al., "Developing Countries Must Lead on Solar Geoengineering Research," *Nature* 556 (2018): 22–24.

CONCLUSION

1. Potential Energy, "Later Is Too Late: A Comprehensive Analysis of the Messaging That Accelerates Climate Action in the G20 and Beyond" (2023), https://potentialenergycoalition.org/later-is-too-late-global-report/.

Index